Springer Series in Optical Sciences   Volume 60

Editor: Theodor Tamir

Springer Series in Optical Sciences

Editorial Board: A. L. Schawlow K. Shimoda A. E. Siegman T. Tamir

Managing Editor: H. K. V. Lotsch

Volumes 1–41 are listed on the back inside cover

Yu. I. Ostrovsky   V. P. Shchepinov
V. V. Yakovlev

Holographic Interferometry in Experimental Mechanics

With 167 Figures

Springer-Verlag

Berlin Heidelberg New York London
Paris Tokyo Hong Kong Barcelona

Professor Yuri I. Ostrovsky

A. F. Ioffe Physico-Technical Institute, Academy of Sciences of the USSR
194021 Leningrad, USSR

Valeri P. Shchepinov, Assistent Professor

Institute of Physical Engineering, 115409 Moscow, USSR

Victor V. Yakovlev, Assistent Professor

Institute of Physical Engineering, 115409 Moscow, USSR

ISBN 3-540-52604-8 Springer-Verlag Berlin Heidelberg New York
ISBN 0-387-52604-8 Springer-Verlag New York Berlin Heidelberg

Library of Congress Cataloging-in-Publication Data. Ostrovsky, Yu. I. (Yuri Isaevich) Holographic interferometry in experimental mechanics / Yu. I. Ostrovsky, V. P. Shchepinov, V. V. Yakovlev. p. cm. – (Spinger series in optical sciences ; v. 60) Translated from the Russian. Includes bibliographical references and index. ISBN 0-387-52604-8 (U.S.) 1. Holographic testing. 2. Holographic interferometry. 3. Deformations (Mechanics) I. Shchepinov, V. P. (Valeri P.), 1947– . II. Vakovlev, V. V. (Victor V.), 1920– .III. Title. IV. Series. TA417.45.087 1990 620.1–dc20 90-40660

This text was prepared using the PS™ Technical Word Processor

54/3140-543210 – Printed on acid-free paper

Preface

This monograph deals with diverse applications of holographic interferometry in experimental solid mechanics.

Holographic interferometry has experienced a development of twenty years. It has enjoyed success and suffered some disappointments mainly due to early overestimation of its potential. At present, development of holographic interferometry is progressing primarily as a technique for quantitative measurements. This is what motivated us to write this book - to analyze the quantitative methods of holographic interferometry.

The fringe patterns obtained in holographic interferometry are graphically descriptive. In the general case, however, because they contain information on the total vectors of displacement for points on the surface of a stressed body, the interpretation of these interferograms is much more complicated than in typical conventional interferometry. In addition, the high sensitivity of the method imposes new requirements on the loading of the objects under study. New approaches to designing loading fixtures are needed in many cases to ensure the desired loading conditions. The wealth of information obtained in holographic interferometry necessitates the use of modern computational mathematics. Therefore, practical implementation of the various methods of holographic interferometry must overcome substantial difficulties requiring adequate knowledge in diverse areas of science such as coherent optics, laser technology, mechanics, and applied mathematics.

Experimental methods play a significant role in solid mechanics. In some cases they are used to check theoretical calculations or to refine mathematical models, while in others they may constitute the only possible way to solve a problem. Conventional experimental methods such as tensometry, crossline screens, moiré patterns, photoelasticity, etc. have apparently exhausted their capabilities, and only minor technical improvements are expected.

The use of holographic interferometry, on the other hand, opens up radically new possibilities for the contactless measurement of surface displacements and deformations of a specimen or a structural element. Moreover, holographic interferometry provides a new approach to traditional methods (moiré contouring, photoelasticity) and substantially broadens their possible applications.

The application of holographic interferometry is hampered by the lack of specialized literature covering the entire range of problems arising in their implementation. The presently available books do not provide adequate coverage of the information necessary for experimental solid mechanics. In particular, in most cases the quantitative aspects of fringe treatment are treated in insufficient detail. A comprehensive analysis of the theory has been made by W. Schumann, J.P. Zürcher and D. Cuche: *Holography and Deformation Analysis*, published as Volume 46 of the present book series. That book does not, however, consider the various problems connected with the practical use of holographic interferometry in experimental mechanics. In contrast, the present treatise focuses attention primarily on the experimental aspects. Therefore, we recommend that readers of this book wishing to gain a deeper understanding of the theory should refer to the monograph by Schumann, Zürcher and Cuche, as a theoretical supplement.

Many questions pertaining to the various uses of holographic interferometry and examples for the solution of concrete problems can be found in a large number of scientific papers. However, access to them is limited for many researchers and engineers. Besides, they should be treated with a certain caution because of differences in the approaches employed by different researchers, of controversial results obtained in some cases, and of lack of adequate information concerning the experimental techniques and methods used. Furthermore, some studies appear to be of purely illustrative nature and have not been brought close to practical application.

These considerations have motivated us to write this book and have determined the selection of the material to be included. In some cases we have been guided not only by ideas of usefulness, but also by our own scientific interests and preferences. This book primarily addresses the research community and engineers engaged in experimental mechanics. It is intended to be a guide to a thorough use of quantitative holographic methods in scientific investigations and practical work. Particular attention is focused on illustrative material. Several holographic experiments were carried out with the specific purpose of illustrating the respective principle.

We gratefully acknowledge the generous cooperation of our colleagues V.S. Pisarev, V.S. Aistov, A.V. Osintsev, I.N. Odintsev, V.V. Balalov, N.O. Reingand, and A.V. Sivokhin in the experimental work. We are deeply indebted to Dr. H. Lotsch and Professor T. Tamir whose active and kind assistance has made the publication of this book possible. Special thanks are due to Dr. G.P. Skrebtsov, who, in the course of translation of the manuscript into English, had to go into details not only of linguistics but also of a physical and technical nature, and in this way has contributed greatly to the clarity and intelligibility of the presentation.

Y.I. Ostrovsky
V.P. Shchepinov
V.V. Yakovlev

Leningrad
March 1990

Contents

1. Introduction to Optical Holography

Holography is a method of wave recording and reconstruction based on freezing the intensity distribution in an interference fringe pattern, called the *hologram*, which is formed by an object wave and a reference wave coherent with it. The hologram, when illuminated with the reference wave, reproduces the same amplitude and phase distribution as that created by the object wave in the recording. Therefore, in accordance with the Huygens-Fresnel principle, the hologram transforms the reference wave into a facsimile of the object wave. This transformation is achieved practically irrespective of the way in which the intensity distribution in the fringe pattern is recorded, i.e. in the form of variations in the absorption or reflection coefficient (*amplitude hologram*) or in the refractive index or relief (*phase hologram*). A hologram can be recorded on a surface (*two-dimensional recording*) or in a volume (*three-dimensional hologram*). While many characteristics of the wave recording and reconstruction process in allthese cases are different, the principal property of the hologram - the ability to transform a reference wave into a facsimile of the object wave - remains the same.

The principle of the hologram can best be described by the following simple illustration. Imagine an interference pattern formed by a reference and an object wave, and recorded as a photographic positive in the form of a two-dimensional amplitude hologram. Then the regions of the hologram with maximum transmittance willcorrespond to the parts of the object wavefront where its phase is equal to that of the reference wave. The more transparent these regions are, the higher was the intensity of the object wave. Subsequent illumination of such a hologram with the reference beam willproduce, in its plane, an amplitude and phase distribution coinciding with that of the original object wave, thus ensuring appropriate reconstruction of the latter.

The principles of holography were first set down in 1948 by the British physicist *Gabor* [1.1]. The word "holography" derives from the Greek ὅλοσ meaning "whole". In this way the inventor of holography intended to stress that the hologram preserves complete information on both the phase and amplitude of a wave. *Gabor's* initial goal was to improve the principles of electron microscopy. He proposed to record, besides the amplitude, also the phase information of electron waves by superimposing a coherent reference wave. Because no source of coherent electron waves was then available, he limited himself to carrying out optical experiments which mark the beginning of holography. However, the difficulties stemming from the absence of powerful sources of coherent light were so severe, that for a long time holography remained little more than an optical paradox.

1

Holography was revitalized in 1962-1963 when the American scientists *Leith* and *Upatnieks* [1.2, 3] utilized the newly invented laser, and proposed an off-axis reference-beam configuration for hologram formation [1.4, 5]. The Soviet physicist *Denisyuk* [1.6, 7] produced the first holograms recorded in a three-dimensional medium, thus combining Gabor's idea of holography with Lippmann's method of color photography. This initiated an explosive development of the method. By 1965-1966, the theoretical and experimental foundations of holography were laid down, while in subsequent years progress in this area was primarily confined to improvements in applications. An exposition of the fundamentals of holography and of its various applications can be found in many books [1.8-21] to which we refer the reader for a more detailed account of the background material.

1.1 Fundamentals of Holography

1.1.1 Basic Equations

Figure 1.1a displays the schematics of hologram formation. A photographic plate receives light waves simultaneously from an object and from a reference source. We willassume the complex amplitude vectors of all these waves to be normal to the figure plane and, hence, willrestrict ourselves to a scalar treatment. For the complex amplitudes of the object and reference waves at the hologram plane (X_2, X_3) we can write

$$A_0 = a_0 \exp(-i\phi) , \quad A_r = a_r \exp(-i\psi) , \tag{1.1}$$

where a_0 and a_r, ϕ and ψ are the amplitudes and phases of the object and reference waves, respectively, which represent functions of the coordinates X_2, X_3. If these waves are coherent, they willform an interference pattern on the photographic plate with an intensity distribution given by

$$I(x_2, x_3) = |A_0 + A_r|^2 = (A_0 + A_r)(A_0^* + A_r^*)$$

$$= a_0^2 + a_r^2 + A_0 A_r^* + A_0^* A_r . \tag{1.2}$$

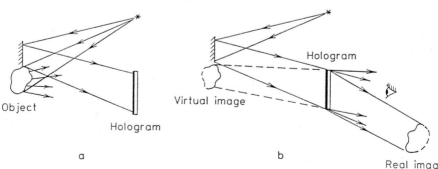

Fig.1.1. (a) Schematic diagram of hologram formation, and (b) wavefront reconstruction

2

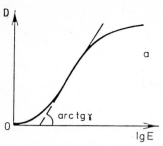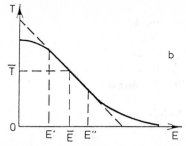

Fig.1.2. (a) Sensitometric curve of photographic emulsion, and (b) plot of amplitude transmittance vs. exposure

We willlimit ourselves to the case of pure amplitude recording this intensity distribution, i.e. where the photographic layer responds to illumination only by changing its transmittance.

The properties of a photographic plate are usually described in terms of the so-called H-D curve (Hurter and Driffield) which represents the dependence of the optical density of the plate, i.e., the quantity $D = \log(1/\tau)$, where τ is the *intensity transmittance* of the developed emulsion layer (Fig.1.2a) on the logarithm of the exposure E (=It, t being the exposure time). The slope of the H-D curve defines the contrast of the photographic material, γ.

For convenience, in describing the holographic process, it is appropriate to represent the properties of a photographic plate in terms of a curve relating the *amplitude transmittance* of the photographic layer $T = \sqrt{\tau}$ to the exposure (Fig.1.2b). Within the range between E' and E" this curve can be approximated by the straight line

$$T = b_0 + b_1 E , \qquad (1.3)$$

with the coefficient b_1 defining the slope of the straight-line section (for negative recording to which Fig.1.2b corresponds, $b_1 < 0$). Substituting into (1.3) the quantity I defined by (1.2), we obtain for the amplitude transmittance distribution of a hologram:

$$T = b_0 + b_1 t(a_0^2 + a_r^2) + b_1 t A_0 A_r^* + b_1 t A_0^* A_r . \qquad (1.4)$$

Illuminating the hologram with the reference wave A_r we find, immediately behind its plane, the following distribution of the complex amplitudes:

$$TA_r = [b_0 + b_1 t(a_0^2 + a_r^2)]A_r + b_1 t a_r^2 A_0 + b_1 t A_r^2 A_0^* . \qquad (1.5)$$

The first term on the right of this equality is equal, within the factor $[b_0 + b_1 t(a_0^2 + a_r^2)]$, to the complex amplitude of the reference wave and corresponds to a zero-order wave. The second term differs from the complex amplitude of the object wave only by the real factor $b_1 t a_r^2$. This term describes the object wave reconstructed by the hologram (positive first-

3

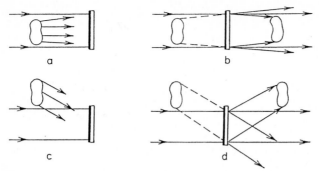

Fig.1.3. Schematic diagram of hologram formation and wavefront reconstruction by (**a, b**) Gabor, and (**c, d**) Leith-Upatnieks

order wave). This wave produces a virtual three-dimensional object image located at the place occupied by the object during the hologram recording (Fig.1.1b). The third term in (1.5) differs from the wave conjugate to the object wave by the complex factor $b_1 t A_r^2$. It describes the negative first-order wave forming a real image of the object. To obtain a nondistorted real object image, the hologram must be illuminated with a wave A_r^* conjugate to the reference wave, i.e., a wave with a wavefront curvature equal in magnitude and opposite in sign to that of the reference but propagating in the opposite direction.

The angles of propagation of the zero and first-order waves are determined by the angles that the object and reference waves make with the hologram plane. In *Gabor's* configuration, the reference source and the object were placed on the hologram axis (in-line arrangement) (Fig.1.3a). Here allthe three waves propagate behind the hologram in the same direction, thus creating mutual interference (Fig.1.3b). In the configuration of *Leith* and *Upatnieks*, this interference was removed by using an inclined reference wave (off-axis configuration) (Fig.1.3c,d).

1.1.2 Hologram Classification

Holograms can be classified according to the method of formation of the object and reference waves and by the way in which the interference structure is recorded. In the first case, the object wave is formed in the following manner. The object is illuminated with a beam of coherent light. The diffracted wave carrying information on the object falls onto the hologram. Depending on the actual position of the object with respect to the hologram and on the optical elements present between them, the relation between the amplitude and phase distributions in the hologram plane and the corresponding distributions immediately behind the object can be described by the Fresnel and Fourier transforms, respectively:

$$a_0(x_2, x_3) = \iint f(x_2', x_3') \frac{1}{\lambda d} \exp\left\{\frac{i\pi}{\lambda d}[(x_2' - x_2)^2 + (x_3' - x_3)^2]\right\} dx_2' dx_3',$$

(1.6)

4

$$a_0(x_2, x_3) = \iint f(x_2', x_3') \exp\left\{ -\frac{2\pi i}{\lambda F}(x_2' x_2 + x_3' x_3) \right\} dx_2' dx_3' , \qquad (1.7)$$

where x_2, x_3 and x_2', x_3' are the coordinates in the hologram and object planes, respectively, d is the distance from the object to the hologram, and F is the focal length of the lens used in the Fourier transformation. If, however, the object lies in the hologram plane or is focussed onto it, then the amplitude and phase distributions on the hologram willcoincide with those in the object plane (Fig. 1.4a). The corresponding recordings are called *image holograms*. In contrast to other methods of hologram formation, in image holograms each point of the object corresponds to a point (small area) on the hologram, thus resulting in a local recording. This has far-reaching consequences. In particular, the spectral composition and the wavefront configuration of the reconstructing wave are no longer important so that there is no need to maintain identical reference and reconstructing waves.

When a hologram is at an infinite distance from the object, i.e., in the Fraunhofer-diffraction region, it is called a *Fraunhofer hologram*. In this case, each point of the object directs a parallel light beam onto the hologram; the relation between the amplitude and the phase distributions of the object wave in the hologram and the object planes are given by the Fourier

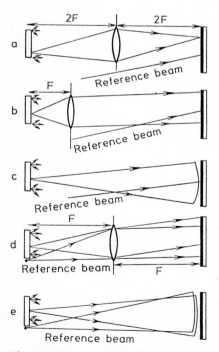

Fig. 1.4. Generation of holograms of various types: **(a)** image, **(b)** Fraunhofer, **(c)** Fresnel, **(d)** Fourier, **(e)** "lensless" Fourier

transform. To produce such a hologram, the object has to lie sufficiently far away from the plate or at the lens focus (Fig.1.4b).

Fresnel holograms belong to the most general class (Fig.1.4c). They are formed when the recording material is placed in the near-field diffraction region (Fresnel-diffraction zone). As the distance between an object and the hologram increases, the hologram transforms into a Fraunhofer hologram. When this distance reduces to zero, we obtain an image hologram.

Since a hologram reproduces the interference pattern created by an object and a reference wave, the shape of the wavefront of the latter is likewise essential for the classification of holograms. If both the object and the point source of the reference wave lie at infinity, the amplitude distribution of each wave in the plane of the hologram coincides with the Fourier transform of the amplitude distribution of the object and the reference source, respectively. Such a hologram is called the *Fourier transform hologram*. Fourier-transform holograms are usually produced by placing the object and the reference source in the focal plane of a lens (Fig.1.4d).

Another configuration for obtaining Fourier-transform holograms is shown in Fig.1.4e. In this case both the object and the reference source are located at a finite distance from the photographic plate. However, since they are equally spaced from the photographic plate, the reference wavefront and the fronts of the elementary wavelets diffracted by the individual points of the object are of equal curvature. Therefore such a *lensless Fourier-transform hologram* has practically the same structure and properties as a hologram produced with the configuration of Fig.1.4d.

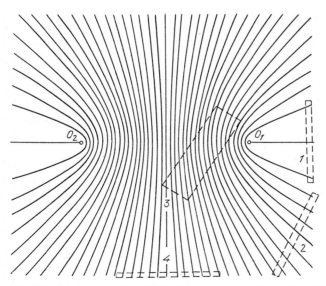

Fig.1.5. Formation of a hologram structure of a point object 0_1 (0_2 is a reference source) by (*1*) Gabor, (*2*) Leith-Upatnieks with off-axis reference beam (*3*) Denisyuk (three-dimensional with opposed-beam recording), (*4*) "lensless" Fourier hologram

The structure of a volume interference pattern created in the superposition of two spherical waves is presented in Fig.1.5. The reference point source O_1 forms the center of one spherical wave, and the point source O_2 that of another wave. The contour surfaces of the maxima and minima in this case represent a system of hyperboloids of revolution. Figure 1.5 illustrates their intersection with the figure plane.

The spatial frequency ν of the interference pattern (and its period d) are determined by the reference-to-object beam angle α at a given point:

$$\nu = 1/d = 2\sin(\alpha/2)/\lambda . \qquad (1.8)$$

The structure is oriented such (Fig.1.5) that the tangent to the surface maximum-intensity contour, at each point of the structure, bisects the angle α, i.e., coincides with the bisector of the interior angle between the wave vectors k_1 and k_2 of the interfering waves. [The wave vector is directed along the direction of propagation (for an isotropic medium this direction is perpendicular to the wavefront surface). The absolute value of the wave vector is $2\pi/\lambda$].

In the in-line arrangement of Gabor the reference source and the object lie on the axis of the hologram. The position of the photographic plate for obtaining an in-line hologram is designated in Fig.1.5 by *1*. In this case the angle α is close to zero and the spatial frequency of the interference pattern is minimum. An *in-line holograms* is sometimes called a *single-beam hololgram* since the Gabor arrangement makes use of one light beam; part of this beam, diffracted by the object, forms an object wave while another part of the original beam, which passes through the object without distortion, acts as a reference wave.

Position *2* in Fig.1.5 corresponds to the arrangement of *Leith* and *Upatnieks* with an off-axis reference beam [1.4]. The coherent reference beam is here produced separately. Therefore the hologram obtained with an off-axis reference-beam configuration is sometimes called a *double-beam hologram*. The spatial frequency of the interference pattern for an off-axis reference beam hologram is higher than that for single-beam hologram. Therefore, recording such holograms requires the use of photographic material capable of high spatial resolution.

Position *3* in Fig.1.5 illustrates to the *opposed-beam* configuration. In this case, the reference and object beams strike the photosensitive layer from opposite sides, the angle between them being close to 180°. The spatial frequency of the structure is maximum and close to $2/\lambda$. In opposed-beam recording the surface of maximum intensity lies in the bulk of the material along its surface. This arrangement was first proposed by *Denisyuk* [1.6, 7]. Since the reconstructed object wave, obtained by illuminating such a hologram with a reference beam, propagates in the direction opposite to that of the illuminating beam, such holograms are also called *reflection holograms*. Figure 1.5 also indicates the formation of a lensless Fourier transform hologram on a photographic plate (Position *4*).

The above classification refers only to the shape of the object and reference waves, and to their mutual orientation which determines the fringe pattern in a hologram. Holograms can be also classified according to the way in which this pattern is recorded. If the thickness of the photosensitive layer is much larger than the spacing between the closest maximum-intensity surfaces, the hologram should be considered as *volume* or *three-dimensional* hologram. The volume properties manifest themselves most clearly in the opposed beam arrangement.

If, however, the holographic pattern is recorded, not in the bulk of the photosensitive layer, but rather on its surface, or if its thickness h is sufficiently smallcompared with the spacing of the neighboring surfaces d, such holograms are called *two-dimensional* or *plane*. The criterion which discriminates between the two- and three-dimension holograms can be expressed as

$$h \geq 1.6d^2/\lambda ,\qquad\qquad(1.9)$$

where λ is the wavelength in the recording medium.

The interference pattern can be recorded in a photosensitive material in either of the following ways:
• in the form of variations in the transmission or reflection index. When reconstructing the wavefront, such holograms modulate the amplitude of the illuminating wave and thus are called *amplitude holograms*,
• in the form of variations in the thickness or refractive index. Such structures are called *phase holograms*.

In many cases phase and amplitude modulations occur simultaneously. For instance, a conventional photographic plate responds to an interference pattern with variations in blackening, refractive index, and relief. After bleaching, such a hologram willpreserve only phase modulation.

The pattern recorded in a hologram can usually be preserved for a long time, so that the processes of recording and of wavefront reconstruction are separated in time. In this case, the hologram is called *stationary*. There exist, however, media (dyes, crystals, metal vapors) which respond virtually instantaneously to illumination through their phase and amplitude characteristics. In this case the hologram willpersist only during the time the object and the reference waves act on the medium. The wavefront reconstruction occurs simultaneously with the recording as a result of the object and the reference waves interacting with the holographic pattern formed by themselves. Such holograms are called *dynamic*.

Holograms recorded in special media acquire the ability of recording and reconstructing not only the amplitude and phase, but also the polarization of the wave [1.22]. New media and techniques have recently been developed which permit holographic recording and reconstruction of a sequence of events on a picosecond time scale [1.23].

1.1.3 Principal Properties of Holograms

1) The principal property of a hologram which distinguishes it from a conventional photograph is that in the latter case, only the intensity distribution of the incident light wave is preserved, whereas a hologram records,

in addition, the phase distribution of the object wave relative to the phase of the reference wave. Information on the amplitude of the object wave remains, as in photographs, recorded on a hologram in the form of contrast in the interference pattern, while the information on phase is contained in the shape and frequency of the interference fringes. Thus, when illuminated with the reference wave, the hologram reconstructs a facsimile of the object wave with allamplitude and phase details.

2) Amplitude holograms are usually recorded on negative photographic material. The properties of such a hologram are identical with those of a positive hologram, namely, to bright (dark) spots of an object correspond bright (dark) spots in the reconstructed image. This is because, as mentioned above, the information on the amplitude of the object wave is contained only in the contrast of the interference pattern. Its distribution is independent of whether a positive or a negative process is used, since the phase of the reconstructed object wave differs by π in the two cases. This cannot be detected visually. This phase difference can manifest itself in some holographic interferometric experiments, however.

3) In cases where the light from each point of an object illuminates the whole surface area of the hologram being recorded, each smallsection of the latter is capable of reconstructing the whole image of the object. Naturally, a smaller section of the hologram willreconstruct a correspondingly smaller section of the wavefront carrying information on the object. As the size of a hologram section decreases, the quality of the reconstructed image degrades.

In the case of image holograms, each point of an object illuminates a corresponding smallsection of the hologram. Therefore, a fragment of such a hologram willbe able to reconstruct only the part of the object corresponding to it. Holograms of transparencies obtained without a diffuser possess the same property.

4) The brightness range reproduced by a photographic plate does not exceed, as a rule, one to two orders of magnitude. Real objects, however, may exhibit much larger brightness differences. A hologram possessing focussing properties uses, for the recording of the brightest parts of an image, the light falling on allof its area and is therefore capable of reproducing brightness gradations extending over five to six decades.

5) If one uses for the reconstruction of the wavefront the reference source fixed with respect to the hologram, then the reconstructed virtual image willcoincide in shape and position with the object. This correspondence breaks down if one changes the position of the reconstructing source relative to the reference source, or its wavelength or the orientation of the hologram or its scale. Such changes are accompanied, as a rule, by aberrations in the reconstructed image.

Here we present expressions which permit calculation of the position of the reconstructed image and its magnification [1.24] for any configuration. We introduce the following notations (Fig.1.6): the hologram lies in plane (x_2, x_3) at $x_1 = 0$ (it is assumed that this position is maintained during both the recording and the reconstruction of the wavefront); the object

9

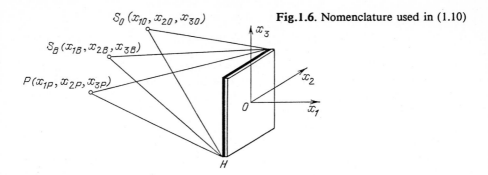

$S_0 (x_{10}, x_{20}, x_{30})$

$S_B (x_{1B}, x_{2B}, x_{3B})$

$P(x_{1P}, x_{2P}, x_{3P})$

Fig.1.6. Nomenclature used in (1.10)

coordinates are x_{1P}, x_{2P}, x_{3P}, the coordinates of the reconstructed image are x_{1r}, x_{2r}, x_{3r}, those of the reference point source x_{10}, x_{20}, x_{30}, and those of the point source used in reconstruction x_{1B}, x_{2B}, x_{3B}. We assume that before reconstruction, the hologram was magnified m times, and that the wavelength of the reconstructing source is μ times longer than that used in recording the hologram. Then

$$x_{1r} = \frac{m^2 x_{10} x_{1B} x_{1P}}{(m^2 x_{10} - \mu x_{1B})x_{1P} + \mu x_{10} x_{1B}} ,$$

$$x_{2r} = \frac{\mu m x_{10} x_{1B} x_{2P} + (m^2 x_{2B} x_{10} - \mu m x_{20} x_{1B})x_{1P}}{(m^2 x_{10} - \mu x_{1B})x_{1P} + \mu x_{10} x_{1B}} , \qquad (1.10)$$

$$x_{3r} = \frac{\mu m x_{10} x_{1B} x_{3P} + (m^2 x_{3B} x_{10} - \mu m x_{30} x_{1B})x_{1P}}{(m^2 x_{10} - \mu x_{1B})x_{1P} + \mu x_{10} x_{1B}} .$$

The angular magnification of a hologram is always, irrespective of the quantities entering (1.10), equal to μ/m. Therefore, for the linear transverse magnification defined as

$$M_\perp = \mu x_{1r}/m x_{1P} , \qquad (1.11)$$

we obtain, in accordance with (1.10),

$$M_\perp = (1 + m^2 x_{1p}/\mu^2 x_{1B} - x_{1P}/x_{10})^{-1} . \qquad (1.12)$$

If a hologram simultaneously produces two images of an object, then the coordinates of the second image can be obtained by substituting $-\mu$ for μ in (1.10 and 12).

Generally speaking, the longitudinal magnification of a hologram differs from the transverse one, i.e.,

$$M_\parallel = \frac{\partial x_{1r}}{\partial x_{1P}} = m^2 \mu (1 + m^2 x_{1P}/\mu x_{1B} - x_{1P}/x_{1B})^{-1} = M_\perp^2/\mu . \qquad (1.13)$$

10

Equations (1.10-13) are not completey exact. They have been derived under the assumption that the hologram-to-object distance is much less than the transverse size of the hologram.

6) Any image of a diffuse object obtained in coherent light, including hologram-reconstructed images, typically exhibits a chaotic granular pattern. This so-called *speckle structure* results from the interference of the light waves reflected from individual elements of the object's microstructure within a spot whose size is determined by the diffraction resolution of the optical system or hologram involved. Thus, the image of an object turns out to consist of individual speckles, in other words, of cigar-shaped blobs of light energy.

The average width of a speckle is

$$\delta l \simeq 1.22\lambda/\alpha , \tag{1.14}$$

where α is the angular diameter of the hologram (or of the lens which forms the image of the diffuse object in coherent light). The longitudinal size of a speckle substantially exceeds its transverse dimension:

$$\delta S \simeq 4\lambda/\alpha^2 . \tag{1.15}$$

Speckles form not only in the image plane, but also at any point in the space surrounding a diffuse object illuminated by coherent light. In this case the speckle size can likewise be calculated with (1.14,15) where α is the angular diameter of the diffusely illuminated surface.

On the one hand, speckles reduce the resolution of holograms as well as of any optical system producing an image of a diffuse (rough) object in coherent light. On the other hand, there are numerous applications of the speckle phenomenon, some of them, similar to the case of holographic interferometry, relate to solid mechanics.

7) The limiting resolution of a hologram is determined by light diffraction at its aperture and can be calculated in the same way as for a conventional optical system. In accordance with Rayleigh's criterion, the angular resolution of a circular hologram of diameter D is

$$\delta\psi = 1.22\lambda/D , \tag{1.16}$$

and for a square hologram of size L,

$$\delta\psi = \lambda/L . \tag{1.17}$$

There are a number of reasons why one cannot technically attain the limiting resolution given by (1.16,17). One of these reasons is the speckle phenomenon. Apart from this, in most configurations the limiting size of the hologram is determined by the resolution of the photographic material, since an increase in hologram size entails an increase of the reference-to-object beam angle and, hence, of the spatial frequency of the interference

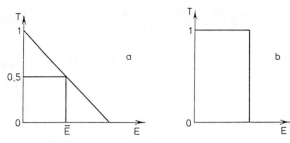

Fig.1.7. Amplitude transmittance T vs. exposure E for (**a**) an ideal linear negative photosensitive material, and (**b**) an extremely nonlinear material

pattern. An exception is the lensless Fourier transform arrangement (Fig. 1.4e) where the spatial frequency does not grow with increasing hologram size.

One should bear in mind that (1.16,17) are valid only in the case where the light from each point on the object is distributed over the entire hologram surface. Otherwise, these expressions should account for the size of the the hologram section which receives the waves from each point of the object.

8) The most essential characteristic of a hologram governing the brightness of the reconstructed image is the *diffraction efficiency*. It represents the ratio of the light flux in the reconstructed wave to that incident on the hologram. The diffraction efficiency is determined by the actual type of hologram and the properties of the photographic material used, as wellas by the recording conditions.

As an illustration we consider an amplitude hologram. As follows from (1.5), the amplitude of the reconstructed wave producing a virtual image is $b_1 t a_r^2 a_0$. Therefore, for the diffraction efficiency we obtain

$$\eta = (b_1 t a_r^2 a_0)^2 / a_r^2 = (b_1 t a_r a_0)^2 . \tag{1.18}$$

Recalling that the contrast p of an interference pattern is determined by the amplitudes of the object and the reference waves

$$p = 2 a_r a_0 / (a_r^2 + a_0^2) , \tag{1.19}$$

we arrive at $a_r a_0 = \frac{1}{2} p (a_r^2 + a_0^2)$. Thus,

$$\eta = \frac{1}{2} b_1 t (a_r^2 + a_0^2) p = (\frac{1}{2} b_1 \bar{E} p)^2 , \tag{1.20}$$

where $\bar{E} = t(a_r^2 + a_0^2)$ is the mean exposure.

Consider an ideal linear, superhigh-sensitivity material with amplitude transmittance vs. exposure characteristic, as shown in Fig.1.7a. In this case $|b_1 \bar{E}| = \frac{1}{2}$, and for a contrast of p = 1 we have, in accordance with (1.20), a diffraction efficiency $\eta_{max} = 1/16 = 6.25\%$. For a photographic material of limiting nonlinearity (Fig.1.7b) the maximum diffraction efficiency of a

Table 1.1. Maximum attainable diffraction efficiency of different types of holograms in percent

Hologram type	Transmission Amplitude	Phase	Reflection Amplitude	Phase
Two-dimensional	6.25	33.9	6.25	100
Three-dimensional	3.7	100	7.2	100

hologram is somewhat higher and amount to 10.1%. The maximum attainable diffraction efficiency of holograms of different types is presented in Table 1.1. The experimentally obtained values are close to the figures given in the Table.

9) If the exposures at the maxima and minima of an interference pattern lie substantially beyond the linear portion of the amplitude transmittance vs. exposure dependence (Fig. 1.2b), hologram recording becomes nonlinear. A linearly recorded hologram may be compared with a diffraction grating having a sinusoidal amplitude transmittance distribution which is known not to produce higher than first-order diffraction. In nonlinear recording, a hologram also represents a periodic grating, however, in this case, the amplitude transmittance distribution may differ substantially from a sine function. Besides the zero- and first-order waves, such a grating also generates diffracted waves of higher orders. A nonlinearity in hologram recording manifests itself, however, not only in this aspect, but also in an amplitude distortion of the reconstructed first order waves. The effects of nonlinearity on the first order image is background enhancement, halation, distortion of the relative intensities coming from different points of the object, and in some cases also, the appearance of false images.

The images formed by higher-order diffracted waves represent complex (autoconvolution-type) functions of the original wave function of the object and have little in common with the properties of the object [1.25]. However, in a number of cases (this relates, for instance, to image holograms or holograms of transparencies obtained without diffusers) the higher-order waves may form images. The brightness distribution in these images is, as a rule, strongly distorted - the phase of the nth order image being n times that of the first-order image. This property of nonlinearly recorded holograms finds application in methods aiming at enhancing the sensitivity of the holographic interferometry for transparent phase objects.

10) Volume holograms also possess unique properties. Such holograms represent essentially a three-dimensional pattern where the contour of maximum and minimum intensities are recorded in the form of variations of the refractive index or reflection coefficient in the medium. When illuminated by a reference wave, such a pattern acts as a 3D diffraction grating (Fig. 1.8). The light reflected specularly from the layers will reconstruct the object wave. Indeed, the maximum and minimum intensity contour (Sect.

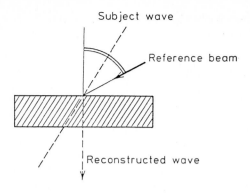

Subject wave

Reference beam

Reconstructed wave

Fig.1.8. Recording of three-dimensional hologram and light diffraction on its structure

1.1.2) bisect the angle made by the object and the reference beams, and this is what accounts for the property of 3D holograms.

The beams reflected from different layers willenhance one another only if they are in phase, i.e., when the path difference between them is λ. This Lippmann-Bragg condition willautomatically be met only for the wavelength at which the hologram was recorded. As a result, the hologram willbe selective with respect to the wavelength of the source used to reconstruct a wavefront. It is thus possible to reconstruct an image by using a continuum source (incandescent lamp or the Sun). If a hologram was exposed to light of several spectral lines (e.g., blue, green, and red), then each wavelength willform its own 3D pattern. When illuminating the hologram, the corresponding wavelengths willbe isolated from the continuum, thus resulting in a reconstruction not only of the pattern, but of the spectral composition of the light wave as well, i.e., it produces a color image.

Three-dimensional holograms produce only one image at a time (virtual or real, depending on the way in which they are illuminated) and do not generate zero-order waves.

1.2 Techniques of Experimental Holography

1.2.1 Holographic Equipment

The holographic equipment used presently can be divided into two groups, namely stationary laboratory installations and special purpose apparatuses which permit experiments on operating machines or test benches.

The most important piece of a holographic equipment is the vibration-isolation platform on which the optical elements and the object to be studied are mounted. The platforms used in mechanical research are made, as a rule, of metal and are provided on the operating surface with a set of holes or grooves for fixing the components of the apparatus and of the object under study. Probably the best option for the platform is the honeycomb design which combines rigidity, vibrational stability, and modest weight. Such platforms are made of steel and aluminum, and may reach 10 m in length, a dimension which is practically impossible to obtain with any other method of platform manufacture. Universal honeycomb blocks produced by Newport Corporation permit assembly of platforms for any shape of holographic systems.

The holographic platform has to be vibration-isolated for the following reason. As already mentioned, the position of the fringes recorded on a hologram is determined by the phases of the beams interfering in the plane of the recording. A change of the beam phase difference by about π during exposure willresult in a washout of the pattern and a dramatic degradation of the hologram quality, resulting in a possible disappearance of the reconstructed image. Changes in the relative phases between the beams may originate not only from variations in the optical properties and the position of the object, but also from displacements and vibrations produced by external mechanical and acoustic factors.

The presence of vibrations in an apparatus can be checked with a microscope mounted in the hologram region [1.26]. One can readily see whether the interference fringes stay fixed by observing their position against an appropriate reference. If the fringe pattern shifts, one should test the mechanical stability of the individual units in the installation. The effect of vibration on the quality of holograms can be eliminated by using industrial shock absorbers of various kinds, for example inflated car and aircraft inner tubes, rubber mats, etc.

The optical systems for hologram construction are made up of various mirrors, lenses, beam splitters, objectives, and other elements which we briefly discuss. The reflecting surface of a mirror should be clean and of good quality so as not to introduce additional distortions into the wavefront. The setup should be provided with a set of spherical and cylindrical lenses of different focal lengths which could be used to illuminate the object under study and to produce reference waves. In quantitative measurements of displacement, the use of collimated radiation (plane wavefront) frequently stems from considerations of convenience in interpreting holographic interferograms (Sect.4.1.1). The collimators used here differ from those of conventional design in that a pinhole aperture is placed at the focus of the first lens. The purpose of this aperture is to spatially filter the light waves diffracted by dust particles and inhomogeneities in the optical system prior to the collimator exit objective. Because the expansion of the beam diameter in the collimator is on the order of 100, alldust particles and inhomogeneities are clearly visible as diffraction rings, which distort the spatial uniformity of the intensity distribution in the beams and change the conditions of exposure at different points on the hologram.

Pinhole aperture *1* (Fig.1.9) is usually 20 to 30 μm in diameter and made of aluminum, nickel, or other foil, 10 to 12 μm thick and positioned

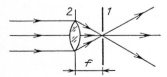

Fig.1.9. Micro-objective lens 2 with pinhole adjustment 1

at the focus of the first lens of collimator 2 (most frequently, a micro-objective). To ensure precise matching of the aperture with the focal point of the micro-objective, the latter is mounted on an adjustable support capable of motion in the direction of the collimator axis, and the aperture is placed on mounts movable in two coordinates transverse to the collimator axis. Obviously, the pinhole position must be adjusted most carefully in both the longitudinal and transverse directions, otherwise a large fraction of the radiation energy willbe blocked. With a correct adjustment, the aperture produces an ideal spherical wave devoid of any aberration present in the optical system forming the beam, of interference caused by secondary reflections, or of diffraction from dust particles and optical imperfections.

The optimum intensity ratio for the reference and object beams determined by the characteristics of the beam splitter, is chosen on grounds of light scattering properties of the object under study and the object-to-hologram distance. The less light scattered from the object toward the hologram, and the larger the distance to the hologram, the higher the part of the light-wave energy that should be injected into the reference beam in order to provide optimum recording conditions at the hologram [1.27]. Experiments suggested an optimum rate of

$$a_0{}^2/a_r{}^2 \simeq 0.3 \ .$$

Deviations (sometimes very substantial) from this requirement can be very useful in holographic studies of vibrations of objects. This subject will be treated in greater detail in Chap.8.

For beam splitters one usually employs plane parallel plates and wedge-shaped glass plates with a dielectric or metallic reflection coating with a transmittance of about 0.3 to 0.7, deposited on one of the faces. Obviously enough, to be usable with different objects, a versatile beam splitter should be capable of readjustment, i.e., it should allow continuous variation of the amplitude ratio of the reference and object beams. A convenient beam splitter design is an uncoated glass wedge, with one of the reflected beams used as the reference. The beam ratio can then be varied over a wide range by properly tilting the wedge since, as is wellknown, the Fresnel reflection coefficient depends substantially on the angle of incidence.

In holographic interferometry one widely uses division of the amplitude and wavefront of the light wave. These techniques can be used simultaneously in designing multihologram interferometers.

Fiber optics provides means to feed light beams along curved guides to any point on the object, including the inner spaces of a construction element.

Both the reference wave and the beam illuminating the object can be guided by means of fiber optics [1.28-36]. The required angular divergence of the beam is obtained by properly adjusting the diameter ratio at the entrance and exit fiber end. In this way temperature-induced deformation, acoustic noise, and vibrations higher than those typical in conventional optics can be avoided. One can also use fiber beam splitters for separating the light into reference and object beams.

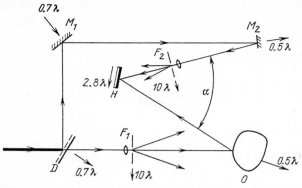

Fig.1.10. Limiting displacement of optical components during hologram recording

We now consider the requirements imposed on the mechanical stability of the individual elements in a typical optical system used for hologram construction. The limiting displacement S, in units of wavelength, of individual elements, which result in a displacement of the interference fringes in the hologram plane by one period during the exposure, are presented in Fig.1.10 [1.37]. The directions of maximum sensitivity with respect to a displacement of the individual elements are specified by arrows. The displacements of the object O and mirror M most seriously affect the hologram quality. Less dramatic is the effect of displacement of the spatial filters F_1 and F_2. A shift in their positions (from the optimum one) results in reduced illumination of the object and hologram. The requirements on temperature stability, when recording holograms with the arrangement in Fig.1.10, have been specified in [1.37]. They are about 0.5°C per meter of light-beam path from the beam splitter D to hologram H. As for elements of the optical system in front of the beam splitter D (including the laser), the requirements imposed on the positional stability during the exposure are not so stringent.

The requirements on the constancy of temperature and humidity, and on the stability of the components of the installation are particularly stringent when three-dimensional reflection holograms are constructed [1.38]. It is known that during the exposure, the relative humidity should be maintained to within fractions of a percent, the temperature to within a few hundredths of a degree, and the holographic installation, particularly the photographic plate, should be isolated against vibration.

The USSR manufactures all-purpose holographic apparatuses SIN-1 [1.39], UIG-2A, UIG-2G-1, UIG-2G-2, and UIG-22 [1.40]. These setups can be used in studies of the deformation of objects under both static and dynamic loading conditions. One should also mention the holographic equipment produced by the Newport Corporation, the Rottencolber Holo System, and the Laboratory of Dr. Steinbichler.

Special holographic setups, frequently called holographic cameras, are playing an increasingly important role in studies of the deformation of structural parts in an operating machine. In cases when strong vibrations or other factors interfere with normal operation of the holographic equipment, pulsed lasers are employed. The use of variable-delay double-pulse lasers makes it possible to record high quality interference patterns of components of operating machines. Descriptions of some of these setups can be found in [1.41-43].

Special techniques have been developed to compensate for the displacements of the objects under study during the hologram construction (some of them willbe discussed in Sect.2.2.4) which permit the use of cw lasers in holographic equipment. Such equipment, mounted on test stands, offers the possibility of recording the specimen displacements, thus providing information on the mechanical properties of materials and areas of stress concention [1.44, 45]. A special setup has been developed to study the deformation of cutting tools [1.46].

Small-size holographic cameras, which are based on the reflex camera and record reflection holograms of focussed images, are used to study the shape of vibrations and to measure displacements [1.47, 48]. An instrument for the construction of holograms of objects and topograms has been used on a manned spacecraft mission [1.49].

Photographing holographic fringe patterns reconstructed by holograms also has its unique features. Already in the stage of selecting an optical setup to record a hologram, one has to bear in mind the following factors significant in subsequently photographiing the fringe patterns:
• the possibility of selecting the desired direction of illumination of the hologram, and the observation of the reconstructed object image;
• the possibility that the zones for a sharp object image and the fringe localization willnot coincide in space, which may require diaphragming the objective lens;
• diaphragming the objective increases the speckle size;
• scale of the reconstructed object image;
• size of the hologram's fragment used in reconstruction.
The fringes should preferably be photographed with a reflex camera with a set of objective lenses on fine-grain films.

1.2.2 Lasers

As already pointed out, holography places particular emphasis on the coherence characteristics of the radiation source. The radiation should possess:
• a sufficiently long coherence time, in order to ensure a high contrast interference pattern in the hologram plane for allparts of the object irrespective of their distance from the hologram;
• a sufficiently high spatial coherence such that the object wave diffracted to each point on the hologram from different parts of the surface under study willproduce a clear interference pattern with any element of the reference wavefront;
• a power sufficiently high to record an interference pattern in a given photosensitive material in the time allowed by the stability of the holographic setup.

The first two conditions can be expressed mathematically, since they follow from the relation between the contrast p of an interference pattern formed in the hologram plane (fringe visibility) and the degree of mutual coherence $|\mu|$ of the object and reference waves [1.50].

$$p = \frac{|\mu| 2\sqrt{\alpha}}{\alpha + 1} , \qquad (1.21)$$

where α is the intensity ratio of the interfering waves. As seen from (1.21), the coherence properties of a source determine the contrast of the pattern recorded on a hologram and, hence, the holographic quality. An analysis of the coherence properties of the various sources demonstrates convincingly that only lasers meet allthese requirements. Without going into the principles of laser operation, we consider briefly the coherence characteristics of the laser radiation.

As is wellknown, a laser cavity supports TEM (transverse electromagnetic) standing waves. The oscillation mode can be characterized by three indices m, n, q, and is denoted TEM_{mnq}.

The first two indices (m, n) refer to the field distribution in the plane normal to the cavity axis. The modes differing in these indices are called transverse. If a laser generates only one of them, the operation is said to be *single mode*. The actual number of simultaneously oscillating modes is determined by the cavity configuration and the way in which each mode couples with the active medium.

The radiation corresponding to the lowest-order transverse mode, TEM_{00q}, concentes close to the cavity axis. The angular divergence of radiation for this mode is minimum and determined by diffraction. An increase in the number of transverse modes results in increasing divergence of the radiation and is equivalent to an increase of the source size.

The mode index q is equal to the number of standing waves which can fit into the cavity length. The oscillation types differing in this index are called longitudinal, or axial, modes. The operation in which a single-mode laser generates only one longitudinal mode is called *single frequency*. The index q is a number much larger than m and n, and is usually omitted in mode designation. Obviously, the coherence length is maximum when the laser operates in the single-frequency regime.

An insufficiently long coherence length limits the depth of the scene being holographed and makes it necessary to equalize the paths of the reference and object beams from the point of their splitting to the hologram. In this case one should also choose an appropriate location of the object relative to the light source and the hologram such that the source-object-hologram paths willbe about equal for allpoints of the object.

Figure 1.11 illustrates the considerations underlying the selection of the most appropriate object location [1.51]. We place the source illuminating the object at point O_1 and the center of the hologram at O_2. The locus of the points for which the sum $B = \Delta_1 + \Delta_2$ of distances from O_1 and from O_2 is a constant, represents the surface of an ellipsoid of revolution with the foci at O_1 and O_2. Figure 1.11 displays the intersection of a family of such ellipsoids for different values of B, the difference in B between the adjacent ellipsoids being $\ell/2$, where ℓ is the coherence length. Obviously, if the il-

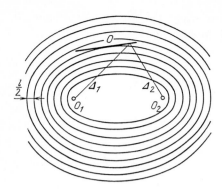

Fig.1.11. Recommendations for appropriate object illumination in the case of limited coherence length in the light source

luminated surface of an object (O) is located such that it willfit between two adjacent ellipsoids, the temporal coherence of the source willbe sufficient for obtaining a hologram of this surface.

CW gas lasers are appropriate for most practical purposes involving the use of holographic interferometry. This type of laser is presently the most widespread in holography for the following reasons:
• a variety of gas lasers are commercially produced, they are readily available and cheaper than other laser types;
• these lasers possess a sufficiently long lifetimes with the major output parameters remaining constant;
• gas lasers are used almost exclusively in reconstructing images from holograms even in cases where the latter have been produced with other types of lasers, as wellas for the adjustment and alignment of interferometers.

According to the composition of the mixture sustaining the gas discharge, gas lasers can be divided into helium-neon (He-Ne), helium-cadmium (He-Cd), argon (Ar), krypton (Kr), and so on. The parameters of some lasers manufactured in other countries can be found in Table 1.2. He-Ne lasers with short discharge tubes up to 50 cm in length operate practically in a single-frequency mode. The shortcoming of such lasers is a low output power (10-15mW). While helium-neon lasers with a resonator length of about two meters operate at a power level of 50-100 mW, their coherence length is usually small(10-20cm). Most of the ion lasers using an active medium of Ar, Kr, or a mixture of both, provide the possibility of mounting a Fabry-Perot etalon inside the cavity so that single-frequency operation can be achieved. In this case the coherence length may reach a few meters, eliminating the need of path equalization for the object and reference beams.

For the investigation of fast processes by holographic interferometry, most promising are pulsed lasers with a solid active medium. Ruby and neodymium lasers are presently the most widespread devices in this area, the latter being ordinarily used in combination with a frequency doubler which converts 1.06 μm emission to 0.53 μm light [1.52].

When using pulsed lasers for hologram recording one should bear in mind that for photographing reconstructed interferograms, it is most convenient to use cw helium-neon lasers (λ = 0.63μm). They may, however,

Table 1.2. Principal charateristics of some gas lasers

Laser type	Active medium	λ^a [μm]	Output power [mW]	Operation
Spectra Physics (USA)				
Model 125	He–Ne	0.6328	50	single-mode, single-frequency
Model 124	He–Ne	0.6328	15	single-mode
Carl Zeiss (GDR)				
HNA-188	He–Ne	0.6328	25	single-mode, single-frequency
Spectra Physics (USA)				
Model 165-09	Ar	0.5145	2000	single-mode, single-frequency
Model 171-09	Ar	0.5145	7500	single-mode, single-frequency
Model 171-01	Kr	0.6471	3500	single-mode; single-frequency
Coherent (USA)				
Inova-20	Ar	0.5145	8300	single-mode, single-frequency
Inova-100-K3	Kr	0.6471	3500	single-mode, single-frequency
Carl Zeiss (GDR)				
ILA-120	Ar	0.5145	700	single-mode, single-frequency
ILK-120	Kr	0.6471	500	single-mode, single-frequency

a The wavelengths specified here are the ones most widely used in holography

lead to distortions associated with the change of wavelength. These distortions are minimum for image holograms. The use of pulsed lasers and, particularly, of twin-pulse lasers substantially alleviates the requirements imposed on the vibration stability of the equipment employed. There are twin-pulse, single-mode lasers manufactured commercially for the need of holographic interferometry. Among them are Model 22 of Apollo Lasers (USA) (pulse energy up to 2.5J, pulse duration 5 to 125ns, pulse interval 1 to 500μs) and those of the Laboratory of Dr. Steinbichler (FRG) (energy up to 0.6J, pulse duration 30ns, pulse interval 1 to 800μs).

Table 1.3. Characteristics of holographic emulsions

Type	Characteristics		
	Wavelength $\lambda[\mu m]$	Sensitivity [erg/cm^2]	Resolution [lines/mm]
USSR:			
PE-2(PFG-03)	0.6328	50000	10000
LOI-2	0.6328	500	10000
VRL	0.6328	50-100	2800
VRP	0.5145	0.02 ASA units	1700
Agfa:			
10E75	0.6328		
	0.6470	20	2800
	0.6940		
8E75	0.6328	75	
	0.6470	75	5000
	0.6940	50	
10E56	0.4880	30	2800
	0.5145	20	
Kodak:			
649F	0.6328	1000	5000

1.2.3 Recording Media

The requirements imposed by holographic interferometry on the recording media may appear modest compared to other areas where holography is employed, namely, the spatial frequency range which should be transmitted by a hologram only rarely exceeds 1500 to 2000 lines/mm (with the exception of reflection holograms), and the noise characteristics and nonlinear distortions are not very important. Therefore, the requirements on the recording media reduce to providing the maximum sensitivity (at the operating laser wavelength) at moderate resolution, noise level and diffraction efficiency. This formulation of the criterion on applicability of a recording medium for holographic interferometry may appear too soft. However, it does reflect the present status of development of recording media for the purpose of holography, an area where it does not seem possible to satisfy allrequirements placed on photosensitive materials.

Indeed, as willbe seen in Sect.2.3, the existing methods of interpretation of holographic interferograms are based on a linear reconstruction of the interference pattern at a low hologram noise level and sufficiently high diffraction efficiency. Much more essential, however, is the problem of broadening the areas of application of holographic interferometry, and the transition from laboratory setups to commercial equipment operating in an industrial environment. This requires shortening the time needed for exposure and processing of a hologram, which also severely restricts the scope of materials that can be used in holographic interferometry. Only two types of materials can actually be applied, namely, silver halide photographic emulsions (plates and films) and photothermoplastics.

Table 1.3 lists the main characteristics of photographic materials manufactured in the USSR and other countries, and are employed in holographic interferometry. Note particularly PE-2 (PFG-03) photographic plates permitting construction of reflection holograms with a high diffraction efficiency. These plates should be processed preferably in a rapid developer bath [1.53], no fixation being required. More detailed information on these materials and their processing can be found in specialized literature [1.54].

A photothermoplastic material essentially consits of a solid solution of abietic acid with an added sensitizer in an amorphous thermoplastic polymer matrix [1.55]. The resolution of these materials usually does not exceed 1000 lines/mm and the sensitivity is hardly inferior to silver halides. Photo Thermo Plastic Carriers (PTPC) are presently used in studies of vibrations and surface relief and in nondestructive testing. Certain progress has recently been made in the use of photorefractive crystals as a recording medium [1.56].

2. Holographic Interferometry

The subject of holographic interferometry encompases construction, observation, and interpretation of interference patterns of waves of which at least one has been recorded on and reconstructed by a hologram. The principle of holographic interferometry is illustrated by Fig.1.1 which displays a general arrangement for obtaining a hologram and reconstructing the wave recorded in it. As has been shown earlier, the interaction of the reconstructing wave with the pattern recorded in the hologram (Fig.1.1b) results in a reconstruction of the object wave. If the reconstructing wave is an exact copy of the reference wave, then both phase and amplitude of the object wave will be faithfully reproduced. If we illuminate the hologram after removing the object, we will see the object image at the exact place and in the same state it occupied during hologram recording. If the object is not removed, then two waves will propagate simultaneously behind the hologram, namely, the wave reconstructed by the hologram, and another one diffracted directly by the object. These waves are coherent and can thus interfere; the interference pattern characterizes the changes which have occurred in the object in the time between hologram recording and observation of the interference pattern. If the state of the object has changed during the observation, for instance, as a result of a deformation or displacement, the interference pattern will have changed, too. Therefore, this method of holographic interferometry is called *real-time interferometry*.

In another variant of holographic interferometry, two (or several) holograms corresponding to different states of the same object are successively recorded on a photographic plate. In simultaneous reconstruction, the waves representing facsimiles of the object waves which existed at different times will interfere. This is the so-called *double (multi)- exposure method*. Sandwich holography, involving the recording of each of the interfering waves on a separate hologram, may be considered as a modification of the double-exposure method.

The extreme case of the multi-exposure method is the *time-average technique* where the hologram of a time-varying (e.g., strained, vibrating or translated) object is exposed continuously. In this case the hologram will record the waves scattered by the object in all the intermediate states it passes during the exposure. The waves reconstructed by such a hologram form an interference pattern that shows how various points of the object displace during its motion.

In conventional as well as holographic interferometry one compares phase contours of two (or several) waves. In conventional interferometry, the waves to be compared are formed simultaneously but propagate along different paths. The time delay between these waves caused by a difference in their optical paths should not exceed the coherence time, and the optical trajectories along which they propagate should be identical, otherwise the interference pattern obtained will describe not only the object under study but the difference in the shape of optical elements in different interferometer arms as well.

Holographic interferometry makes use of the interference of waves passing along the same trajectories but at different instants in time. The obtained interference pattern reflects only the changes in the object, which have occurred between the time of hologram recording and its observation (or in the time between the first and second exposure) and is uniquely related to these changes.

Thus, holographic interferometry is actually a differential method and can be used to compare consecutive states of the same object. The object wave recorded on a hologram, and subsequently reconstructed by it, characterizes the structure of the object down to the tiniest detail. Therefore, holographic interferometry can be employed to study objects of arbitrary shape and even rough, diffusely reflecting surfaces. The only requirement is that in a transition from one state to another, the microstructure of the object should not undergo substantial changes. In contrast, in conventional interferometry the comparison wave can reproduce all details of the object wave only if the latter is of a sufficiently simple shape. Thus, with this method one can study only objects of the simplest shape having a polished optical surface.

The differential nature of holographic interferometry permits one to substantially weaken the requirements on the quality of the optics involved, since the waves that are compared propagate along the same trajectories and are equally distorted by imperfections in the optical elements. In other words, the presence of defects in optical elements in no way affects the interference pattern. This insensitivity to optical defects enables one to study objects of practically unlimited size, whereas in conventional interferometry such investigations are made difficult by the complexity of the arrangement and the high cost of interferometers with large and high quality mirrors.

If a hologram stores an object wave recorded within a large solid angle, then it can be used to reconstruct the interference pattern of the light waves scattered by the object in different directions, which is employed in studies of the deformation of bodies of complex shape.

Holographic interferometry offers the possibility of obtaining interference patterns formed by light waves of different frequencies. For this purpose the photographic plate is exposed to light of a multi-frequency source. Illuminating such a hologram reconstructs the waves of different frequencies recorded on it, which can interfere, since they are reconstructed by the same monochromatic light beam. The multi-frequency method is used, for instance, to change the sensitivity of holographic interferometry as well as to study surface relief.

Most of the holographic-interferometry techniques deal only with the shape of the fringes in an interference pattern. However, the fringe contrast and the position of the fringe localization region are likewise related to the changes the object undergoes. Indeed, fringe contrast gives an idea of degradation in the microstructure of the surface under investigation (due, e.g., to corrosion or wear), while from the fringe localization parameters one can judge the displacement and deformation of the object.

One of the major applications of holographic interferometry is the determination of the displacement of points on the surface of a strained body. In contrast to conventional interferometry, the holographic record bears information on the total displacement vectors of points on the surface of a body. Interpretation of the fringes, i.e., determination, from their arrangement, of the displacement vector for an arbitrary point on the surface under investigation, is achieved by analyzing the fringe patterns obtained from different viewing directions, or by constructing several holograms (the absolute fringe-order technique), or by modifications of these methods (Chap.2). By applying statistical and nonstatistical criteria to the experimental design, one can construct interferometers capable of measuring the displacement-vector components with minimal error (Chap.3). Application of special optical systems enables the study of the field of individual displacement vector components (Chap.4). From holographic interference patterns one can, in some cases, directly determine the derivatives of displacement and, hence, the strains (Chap.5). Holographic interferometry is capable of determining not only elastic, but residual displacements as well (Chap.6).

The technique can also be used to generate contour maps of a surface (Chap.7) with the two-frequency, immersion, or two-source techniques. The first method involves recording a hologram of a surface with a two-frequency light source. In the immersion technique, the object is placed into a cell with a plane window, and two exposures are made in different immersion media (liquid or gas) with different refractive indices. In the two-source method, the hologram is likewise exposed twice but at different directions of the object-illuminating beam. The methods of holographic interferometry also permit a direct comparison of the reliefs of the surface under study and a reference surface. Holographic topography makes possible the measurement of large deformations in objects of complex shape.

Vibrations of objects (Chap.8) are investigated, as a rule, by the time-average technique, in which the hologram of an object is exposed over several vibrational periods. The fringe intensity drops rapidly with the amplitude of vibration. The brightest fringes correspond to the nodal lines. From such interference patterns one can derive the vibration-amplitude distribution over the object surface. The measured amplitude range can be extended by using the stroboscopic method where a hologram is exposed not continuously, but rather at only certain instants of time synchronized with the chosen vibration phase. Here the fringe brightness does practically not depend on the vibration amplitude.

Flaw inspection by interferometric holography is in widespread use and different modifications of the method are being developed. The regular interference pattern produced under loading of an object to be studied becomes distorted at flaws (cracks, cavities, disbonded areas in laminar structures). Indeed, when a crack appears at the surface of a body, interference fringes exhibit a break at its opposite edges. To locate flaws with a fringe pattern, the object is loaded either statically or vibrationally. In some cases the object is locally heated or cooled.

Of particular interest is the investigation of phase objects, such as shock waves in gases and liquids, plasma, or plane transparent models of strained bodies. When they are probed with an object beam, the interferometric hologram shows the spatial distribution of the refractive index, which is uniquely related to that of the atom, molecule, and electron concentration in the volume under study.

These are the principal features and the most essential applications of holographic interferometry.

As mentioned, one of the most important applications of holographic interferometry is the examination of diffusely reflecting objects. The most difficult problem in this task is the formation of interference fringes. There are many different approaches to the description of this phenomenon. In our opinion, the simplest and most illustrative way which reflects nearly all the major features of interference fringes is the geometric approach [2.1] based on ray optics and the calculation of path differences, which we follow in this book. For a more detailed study of the process of fringe formation we refer the reader to the book of *Schumann* and *Dubas* [2.2].

Difficulties in interpretation of the fringe patterns unfortunately prevent broader application of holographic interferometry. Indeed, a holographic interference pattern contains copious information on the total displacement vectors of points on the surface of an object. As already mentioned, the interpretation of interference patterns is based on an analysis of the various characteristics of the fringes, namely, their frequency, orientation, visibility, and localization. While not all of these have found broad application, we believe that the rich variety of problems in mechanics will eventually lead to the recognition of their importance.

Determination of the displacements of points on the surface of a strained object is generally a tedious process. Thus, the question of automation naturally arises. Automated systems of holographic-interferometric measurements are presently employed only in some stages of processing and are not yet in widespread use. This is an independent and very complex problem which is not considered here.

2.1 Formation of Holographic Fringe Patterns

2.1.1 The Two-Exposure Method

The two-exposure method proposed in [2.3] is the most widely used way of constructing holographic fringe patterns which are instrumental to the quantitative determination of surface displacements on a strained body. The method consists essentially of taking a consecutive exposure on the same photosensitive material of two holograms of an object in difference states, an original and a changed state, for instance, prior to and after a loading. When a doubly-exposed hologram is illuminated with a facsimile of the

27

reference wave, both waves scattered by the surface of the body in its two states will be reconstructed simultaneously. Their interference produces an image of the surface of the body superimposed with a system of interference fringes which contain information on the changes the body has undergone between the two exposures. Thus, by using one doubly-exposed hologram the interference of light waves shifted in time can be observed.

We now consider the formation of a fringe pattern in the two-exposure method in more detail. The complex amplitudes of the light waves reflected from the surface of a body in its two states in the hologram plane A_1 and A_2 are written in the form

$$A_1 = a_o \exp(-i\phi) \, , \tag{2.1}$$

$$A_2 = a_o \exp[-i(\phi+\delta)] \, , \tag{2.2}$$

where a_o and ϕ are the amplitude and phase of the wave scattered by the object in its original state, and δ is the strain-induced phase change of the object wave in the hologram plane. The deformation of the body is assumed to affect only the phase of the object wave. In addition, we will assume that the changes in the microrelief of the surface between the exposures may be neglected.

The expression for the complex amplitude of the reference wave A_r can be written by analogy to that for the object wave as

$$A_r = a_r \exp(-i\varphi) \, , \tag{2.3}$$

where a_r and φ are the amplitude and phase of the reference wave, respectively. The total exposure E in the plane of the recording medium, e.g., a photographic plate, after the first and second exposures will be

$$E = |A_1 + A_r|^2 \tau_1 + |A_2 + A_r|^2 \tau_2 \, , \tag{2.4}$$

where τ_1 and τ_2 are the durations of the first and second exposures, respectively. Taking $\tau_1 = \tau_2 = \tau$ and substituting into (2.4) the expressions (2.1, 2 and 3) we obtain, after some algebraic manipulation

$$E = 2(a_o^2 + a_r^2)\tau + \tau a_o a_r \exp(i\varphi)\{\exp(-i\phi) + \exp[-i(\phi+\delta)]\}$$
$$+ \tau a_o a_r \exp(-i\varphi)\{\exp(i\phi) + \exp[i(\phi+\delta)]\} \, . \tag{2.5}$$

We assume that the photographic plate is processed in such a way that its amplitude transmittance T depends linearly on the exposure E

$$T = b_0 + b_1 E \, , \tag{2.6}$$

where b_0 and b_1 are constants characterizing the T vs. E curve (Sect. 1.1.1)

When the doubly-exposed hologram is illuminated with a replica of the reference wave, a wave will form at the back side of the hologram, whose complex amplitude A is proportional to the transmittance:

$$A = Ta_r exp(-i\varphi) \ . \tag{2.7}$$

Substituting (2.5,6) into this equation we obtain, after simple transformations

$$A = [b_0 + 2b_1\tau(a_o{}^2 + a_r{}^2)]a_r exp(-i\varphi) + b_1\tau a_o a_r{}^2\{exp(-i\phi) + exp[-i(\phi+\delta)]\}$$
$$+ \ b_1\tau a_o a_r{}^2\{exp(i\phi) + exp[i(\phi+\delta)]\}exp(-i2\varphi) \ . \tag{2.8}$$

The first term in (2.8) represents, to within a constant factor of $b_0 + 2b_1\tau(a_o{}^2 + a_r{}^2)$, the amplitude of the reference wave traversing the hologram and corresponds to a zero-order wave. The second term describes, to within a real factor of $b_1\tau a_o a_r{}^2$, two object waves forming virtual images. Finally, the third term describes two distorted real images. It should be recalled that when the hologram is illuminated with a wave conjugate to the reference, whose complex amplitude is $a_r exp(i\varphi)$, a nondistorted real image of the body is obvserved to coincide with the body itself. The intensity distribution I_M in the virtual image will correspond to the square of the second term in (2.8):

$$I_M = b_1{}^2\tau^2 a_o{}^2 a_r{}^4|exp(-i\phi) + exp[-i(\phi+\delta)]|^2$$
$$= 2b_1{}^2\tau^2 a_o{}^2 a_r{}^4(1 + cos\delta) = 4b_1{}^2\tau^2 a_o{}^2 a_r{}^4 cos^2(\tfrac{1}{2}\delta) \ . \tag{2.9}$$

This expression may be conveniently rewritten as

$$I_M \sim a_o{}^2 cos^2 \tfrac{1}{2}\delta = I_o cos^2 \tfrac{1}{2}\delta \ , \tag{2.10}$$

where I_o is the intensity of the image of the object in the original state. As seen from (2.10), the intensity of the reconstructed image varies as the cosine squared. In other words, the image of the surface of the body will be superimposed by a system of interference fringes. The pattern of the interference fringes depends on the function δ. For fixed points on the surface $\delta \equiv 0$, the corresponding intensity is at a maximum. In the following discussion we shall call such fringes *zero-motion fringes*. Thus, in two-exposure patterns the zero-motion fringes will be bright.

The *contrast* or *visibility* V of fringes in a double-exposure holographic interference pattern for equal exposure times, which is determined as

$$V = \frac{I_{max} - I_{min}}{I_{max} + I_{min}} \ ,$$

is equal to unity. As follows from (2.10), the fringe pattern does not depend on the sign of the phase difference δ.

When reconstructing light waves recorded on a double-exposed hologram, the shape of the reconstructing beam may sometimes differ substantially from that of the original reference beam. In this case scale distortions and image displacements turn out to be the same for the two waves, how-

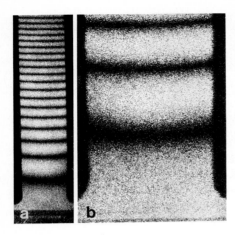

Fig.2.1. (a) Two-exposure holographic interferogram of a cantilever beam subjected to a bending stress, and (b) its magnified fragment

ever, so that the fringe pattern will remain practically unchanged. Likewise, aberrations are the same for both waves and do not lead to noticeable changes in the structure of the fringe pattern. The two-exposure method produces fringes of infinite width. To make the fringe width finite, the object-to-reference beam angle should be changed between exposures.

Figure 2.1a displays a typical fringe pattern obtained by the two-exposure method on a cantilever beam loaded with a concentrated force. The arrangement used was that of Leith-Upatnieks where the reference and object beams are incident on the same side of the holographic plate. Figure 2.1b shows a magnified part of the fringe pattern of the beam near the rigid fixing, which clearly exhibits a speckle pattern. This noise hinders observation and analysis of the fringes, particularly in areas where they are crowded. For reliable resolution of the fringes, their period should be greater than the scale size of the speckle [2.4]. Note that there is no speckle in the images reconstructed by illuminating image and reflection (Denisyuk) holograms.

2.1.2 Sandwich Holography

The light waves reflected from the surface of a strained body in its two states can also be recorded on separarate holograms and the interference fringes observed in their simultaneous reconstruction. This method of forming holograms, first proposed in [2.5,6] to study air flows and subsequently developed by *Abramson* for certain problems in mechanics [2.7,8] has been named sandwich-holographic interferometry.

Technical implementation of this method requires a special kinematic fixture for a precise return of the hologram back to the position at which it was recorded. Therefore, prior to considering the method of hologram construction we will dwell, in some detail, on the design of the device. A photograph of the device similar to the one used in [2.9] is depicted in Fig.2.2. The photographic plate *1* rests on its back on three spheres *2* and with two edges on three cylinders *3*. The vertical plate of the device bearing the spheres and the cylinders is tilted at 8°. Pressed by its weight, the photo-

Fig.2.2. Device for precise repositioning of photographic plate in sandwich holography

graphic plate contacts the spheres and the cylinders in the only position it can occupy in the device. When returning the hologram to the device one should make sure that no chips have formed on the backing in contact with the cylinders, otherwise the plate will not return precisely to the original position.

Two photographic plates, one on top of the other, are placed into the device in such a way that their emulsions will face the object under study. After their exposure in the arrangement depicted in Fig.2.3a (S is a point light source illuminating the object O and S_0 is a reference light source) and photographic processing, one obtains two holograms, H_1 and H_2. In a similar way one produces two holograms of the loaded object (Fig.2.3b). If we now place the holograms H_1 and H_4 corresponding to two different states of the object into the device, and illuminate them with the reference wave, then the observer will see a virtual image of the object (O) with superimposed interference fringes (Fig.2.3c). Exactly the same pattern will be observed if one uses the holograms H_3 and H_2. Each of the sandwich holograms H_1, H_4 or H_3, H_2 thus constructed is equivalent to one doubly-exposed hologram.

It should be borne in mind that implementation of this method entails serious technical difficulties. In the reconstruction of the light waves recorded in sandwich holograms H_1, H_4 (Fig.2.3c), the wave from H_1 passes through H_4 and, hence, inhomogeneity and differences in the thickness of the backing in H_4, as well as the presence of a gap in the sandwich may introduce distortions into the phase of the wave reconstructed by H_1. If, however, the reconstructing wave from the source S_0 passes through H_4, close to the path of the wave reconstructed by H_1, then it will be subjected to the same distortions. This then compensates for the phase distortions due to H_4. Therefore, for a successful application of sandwich-holographic interferometry, the angle between object and reference beams should be as small as possible. This is usually achieved by increasing the distance between the hologram and the object under study. If we interchange the photographic plates H_1 and H_4, then in the reconstruction a system of circular

31

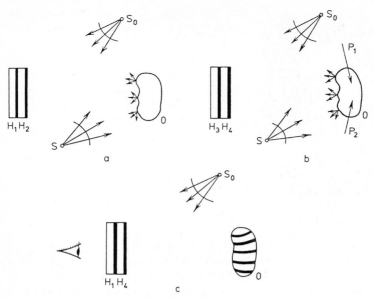

Fig.2.3. Sandwich holographic interferometry: **(a)** recording by $H_1 H_2$ sandwich of undeformed object; **(b)** recording by $H_3 H_4$ sandwich of loaded object; **(c)** wavefront reconstruction by $H_1 H_4$ sandwich

fringes will be observed. The same pattern will be seen if the sandwich is made up by the two first or two second holograms only.

The principal asset of sandwich holography lies in the possibility of recording holograms of the consecutive states that the surface of a stressed object undergoes, allowing an interferometric comparison of any pair of these states. Having obtained the fringe pattern of desired frequency, one can fix the two photographic plates end-to-end to form a sandwich similar to one doubly-exposed hologram. Obviously in this case, the reconstred image will, just as in the case of the two-exposure technique, be modulated by fringes with a cosinusoidal intensity distribution.

2.1.3 Real-Time Method

The real-time holography method involves the interference of waves of which one is reconstructed by means of the hologram of an object recorded in its initial state while the other is scattered directly by the strained object [2.10]. Let us record a hologram of the object under study on a photographic plate. After photographic processing we return it to the location of the exposure. If we now observe the object through the hologram while gradually loading it, interference fringes will become evident and will change their shape and frequency in accordance with the changes in the object. The fringe pattern observed with this real-time technique characterizes the dynamics of the displacement of points on the surface of the object.

32

One can write the expression for the complex amplitude A_1 of a light wave reconstructed by a hologram in the form

$$A_1 = b_1 \tau a_r^2 a_o \exp(-i\phi) , \qquad (2.11)$$

where b_1 is the slope of the T vs E curve, τ is the exposure time, a_r is the reference wave amplitude, and a_o and ϕ are the amplitude and phase of the object wave, respectively. For the complex amplitude of the light wave scattered by the object and passing through the hologram A_2 we write

$$A_2 = \beta a_o \exp\{-i[\phi + \delta(t)]\} , \qquad (2.12)$$

where β is a coefficient accounting for the wave's attenuation in transit through the hologram and $\delta(t)$ is the strain-induced change in the phase of the wave. We assume that during the recording and processing of the hologram until it is returned into the optical system for observation, no changes have occurred in the object under study, i.e., $\delta(t) = 0$. In this case the interfering waves whose complex amplitudes are given by (2.11 and 12) will differ from one another only in the factors $b_1 \tau a_r^2$ and β. If the recording process is positive ($b_1 > 0$), then the interfering waves are in phase and the image of the object should be bright. For negative recording ($b_1 < 0$) the interfering waves will be out of phase and the image will be dark. In other words, with positive processing, the zero-order fringes will be bright, and with negative, dark. For an object under stress the quantity $\delta(t)$ is non-zero, and the image of the surface will be superimposed by interference fringes varying with time.

For the light intensity I in the hologram plane we have

$$I = |A_1 + A_2|^2 = |b_1 \tau a_r^2 a_o \exp(-i\phi) + \beta a_o \exp\{-i[\phi + \delta(t)]\}|^2 . \qquad (2.13)$$

We now assume that the intensities of the object wave reconstructed by the hologram and of the wave reflected from the object are equal:

$$b_1 \tau a_r^2 = \beta . \qquad (2.14)$$

For positive recording ($b_1 > 0$) and taking into account (2.14), Eq.(2.13) can be transformed into

$$I = 2b_1^2 \tau^2 a_r^4 a_o^2 [1 + \cos\delta(t)] = 4b_1^2 \tau^2 a_r^4 a_o^2 \cos^2[\delta(t)/2] . \qquad (2.15)$$

The latter expression can be conveniently rewritten as

$$I \sim a_o^2 \cos^2[\delta(t)/2] = I_0 \cos^2[\delta(t)/2] , \qquad (2.16)$$

where $I_0 = a_o^2$ is the intensity distribution at the surface of a stationary object. Thus, the intensity of the object image under real-time positive recording is modulated by cosinusoidal fringes, just as is the case in the two-exposure method.

In the more conventional case of negative recording ($b_1 < 0$), the intensity of the object image will be distributed as

$$I = 2b_1{}^2 \tau^2 a_r{}^4 a_o{}^2 [1 - \cos\delta(t)] = 4b_1{}^2 \tau^2 a_r{}^4 a_o{}^2 \sin^2[\delta(t)/2] \qquad (2.17)$$

or, reduced to a simpler form,

$$I \sim a_o{}^2 \sin^2[\delta(t)/2] = I_0 \sin^2[\delta(t)/2] . \qquad (2.18)$$

The intensity modulation of the image in real-time negative hologram recording is a sine function.

The fringe contrast in the real-time method depends on the intensity ratio of the interfering waves. If the hologram recording conditions specified by (2.14) are satisfied, maximum fringe visibility will be reached. Another way of equalizing the intensities is by introducing attenuating filters into both the object and reference beams.

Application of the real-time method is limited by the stringent requirement that the hologram be returned to the place of exposure to within one tenth of the wavelength of the light used. Thus one has to match, to within a fraction of the spatial period, the pattern formed by the object and reference beams prior to loading with that recorded on the hologram. The exact repositioning of the hologram can be checked by looking for the absence of fringes on the surface of the unstrained object.

For a quantitative determination of the surface displacement in a strained body, silver halide emulsions are the most frequently used photographic media. There are three major processing techniques in real-time holography. The first entails photographic processing of the plate at the place of exposure [2.11-14]. In this case the photographic plate is fixed in a special holder together with a cuvette, where, after the exposure, the solutions for the emulsion processing are consecutively supplied. A second alternative is mounting the photographic plate in a special kinematic fixture permitting its removal from the optical setup and repositioning it to within a fraction of a fringe spacing [2.15, 16]. Under these conditions, the plate is photographically processed outside the holographic setup. One of the possible versions of the kinematic fixture is shown in Fig.2.4. The detachable section *1*, shaped as a ring, is rigidly fixed to holder *2* accommodating photographic plate *3*. Six triangular grooves spaced 60° from one another are machined in the ring. Three steel balls *4* are fixed in the baseplate. The self-centering of the device, which makes precise repositioning of the plate possible, is effected through the weight of the detachable ring and of the plate holder at the points where the balls touch the surface of any of the three grooves. Shown on the left in Fig.2.4 is the assembled fixture with the photographic plate in place. In the course of the photographic processing the plates should be prevented from knocking against the walls of the cuvettes holding the processing solutions. Such knocks may bring about a displacement of the plate in the holder and, as a result, the appearance of parasitic fringes in the image. Depending on the actual task, the design of the kinematic fixture may vary [2.17]. The hologram can also be reposi-

Fig.2.4. Three-support device for real time holography

tioned by means of the device described in Sect.2.1.2 and shown in Fig.2.2.

When working with photographic plates, one should bear in mind that during their processing and drying the emulsion can suffer swelling, shrinkage and deformation. Homogeneous swelling or shrinkage results in a loss of the diffraction efficiency of a hologram. Studies have shown [2.18] that during development fast swelling occurs, while in the fixation step the emulsion shrinks. Thus, to reduce these phenomena it is desirable to shorten the duration of both development and fixation of the plate. This can be achieved by overexposing the emulsion and then underdeveloping it. In this case only a thin surface layer of the emulsion is developed thereby substantially reducing the fixation time. To weaken the effects due to emulsion swelling and shrinkage one may record holograms on PE-2 photographic plates which require a rapid developer bath (Sect.1.2.3). Their development time does not exceed 10 s, and no fixation is needed.

The strongest effect causing the appearance of parasitic fringes in real-time holography comes from inhomogeneous deformation of the emulsion in the course of its processing. This deformation is due to the residual stress generated in the process of deposition of the emulsion onto the glass substrate. A preliminary soaking of a plate in water may provide an efficient means of relieving this stress [2.19].

Apart from photographic plates, photoplastics have been reported as good recording media in real-time holography [2.20-22]. Special cameras permit immediate processing of photoplastic films on a subsecond time scale. This substantially broadens the possibilities of employing the real-time method not only in a laboratory but also in an industrial environment. presently, however, the use of photoplastics is primarily confined to non-destructive testing based on quantitative analysis of holographic fringe patterns.

2.1.4 The Multi-Exposure Method

We now consider the case of consecutive recording, on a single hologram, of several light waves scattered by the surface of a strained body. We assume that the additional phase difference acquired by each subsequent object wave to be constant and equal to δ. In this case we may write for the complex amplitudes of the object waves:

$$A_1 = a_o \exp(-i\phi) \,,$$
$$A_2 = a_o \exp[-i(\phi+\delta)] \,,$$
$$A_3 = a_o \exp[-i(\phi+2\delta)] \,,$$
$$\text{-----------------------}$$
$$A_m = a_o \exp\{-i[\phi + (m-1)\delta\} \,, \tag{2.19}$$

where a_o and ϕ are the amplitude and phase of the object wave. For the irradiance distribution in the hologram plane we have

$$E = (|A_1+A_r|^2 + |A_2+A_r|^2 + |A_3+A_r|^2 + ... + |A_m+A_r|^2)\tau \,. \tag{2.20}$$

Here A_r is the complex amplitude of the reference wave given by (2.3), and τ is the duration of each of the m exposures. In analogy to the two-exposure method, and assuming the recording to be a linear process, we use (2.6 and 7) to find the amplitude A of the light wave reconstructed by the multiply-exposed hologram:

$$A = [b_0 + b_1 m^2 \tau(a_o^2 + a_r^2)]a_r \exp(-i\varphi) + b_1 \tau a_o a_r^2 \sum_{j=0}^{m-1} \exp[-i(\phi + j\delta)]$$

$$+ b_1 \tau a_o a_r^2 \exp(-i2\varphi) \sum_{j=0}^{m-1} \exp[i(\varphi + j\delta)] \,. \tag{2.21}$$

The second term gives, to within a multiplicative constant, the scattered object-wave amplitudes. For the intensity of the virtual image I_M we can write

$$I_M = b_1^2 \tau^2 a_o^2 a_r^4 \left| \sum_{j=0}^{m-1} \exp[-i(\phi+j\delta)] \right|^2 = b_1^2 \tau^2 a_o^2 a_r^4 \left| \frac{1 - \exp(-im\delta)}{1 - \exp(i\delta)} \right|$$

$$= b_1^2 \tau^2 a_o^2 a_r^4 \frac{\sin^2(m\delta/2)}{\sin^2(\delta/2)} \sim I_0 \frac{\sin^2(m\delta/2)}{\sin^2(\delta/2)} \,, \tag{2.22}$$

where $I_0 = a_o^2$. Thus, the intensity of the image reconstructed with a multiply-exposed hologram is modulated by a complex function of the phase difference.

The virtual-image intensity distributions for m = 2, 4 and 7 are presented in Fig.2.5 by curves 1, 2 and 3, respectively. As the number of ex-

36

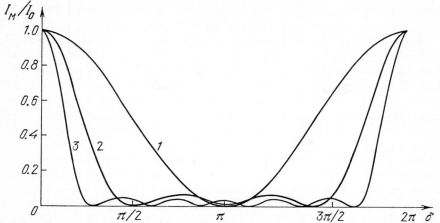

Fig.2.5. Virtual image intensity distribution for different numbers of exposures: (*1*) m=2, (*2*) m=4, (*3*) m=7

posures grows, the principal intensity maxima become narrower. The intermediate maxima between them are of a substantially lower intensity.

To further illustrate this method, Fig.2.6 displays fringe patterns obtained in an opposed beam configuration and corresponding to a deflection of a cantilever beam loaded with a concentrated force. Figure 2.6a shows a fringe pattern recorded by the two-exposure technique for a load increment of 0.2N. The fringe pattern observed in the reconstruction of the image with a quadruply-exposed hologram (Fig.2.6b) (three load increments of 0.2N each) clearly reveals the effect of bright fringe narrowing. As the number of the loading steps increases, this effect becomes enhanced. This is particularly evident from the fringe pattern of the cantilever beam obtained for m = 7 (i.e., six loading steps of 0.2N) presented in Fig.2.6c. As evidenced by an analysis of the interference fringes in Fig.2.6b and c, the displacements of points on the beam's surface corresponding to equal loading

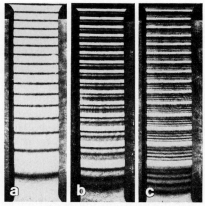

Fig.2.6. Multi-exposure holographic interferogram of the bending of a beam for (**a**) m=2, (**b**) m=4, (**c**) m=7

increments vary from one another at different loading stages by less than $\lambda/2m_1$, where m_1 is the number of loading increments [2.23].

Burch et al. [2.24] were the first to apply this method of holographic interferometry of diffusely reflecting objects to studies of the deformation of a structural element. For a constant incrementation of the loading between the exposures, the narrowing of interference fringes is observed to occur only if the increment of the displacements and the corresponding change in the phase difference are equal, which implies a linear deformation. In cases where the fringe pattern becomes blurred the deformation is nonlinear.

Note that the function given by (2.22) and corresponding to equal exposures of all superimposed holograms is similar to the function describing the intensity distribution in the diffraction of light from a multi-slit aperture [2.12]. The side maxima can be suppressed by creating conditions analogous to those typical of a Fabry-Perot interferometer, namely, by making the amplitudes of the interfering waves not equal but rather decreasing in a geometrical progression. Assuming the recording to be linear, this can be done by changing the amplitudes of the consecutive object waves recorded on a multi-exposure hologram by the same law.

2.1.5 The Time-Average Method

In the above methods the holograms are recorded under conditions where the object is in a steady state. We now consider the case of recording a hologram during a deformation or motion of the object. We limit ourselves here to such motions or deformations where the time function can be isolated [2.25]. In such an approach the change in time of the phase of the object wave in the hologram plane, $\delta(t)$, can be represented as a product of the phase difference δ corresponding to a maximum displacement of the points of the object by a function of time $f(t)$:

$$\delta(t) = \delta f(t) . \tag{2.23}$$

The displacement function for an arbitrary point on the surface of the object, $D(x_1,x_2,x_3,t)$ can be written as

$$D(x_1,x_2,x_3,t) = D_m(x_1,x_2,x_3)f(t) ,$$

where x_1,x_2,x_3 are the coordinates of the point of interest on the object surface, and $D_m(x_1,x_2,x_3)$ is the maximum displacement of this point.

Assume the function $f(t)$ to be the same for all points on the surface. Then for the complex amplitude of the object wave $A(t)$ in the hologram plane at some instant in time one can write

$$A(t) = a_o \exp[-i\delta f(t)] . \tag{2.24}$$

The exposure time of the hologram is denoted by τ. When the hologram is illuminated by the reference wave, the complex amplitude of the reconstructed light wave will be proportional to the time-averaged amplitude of the object wave A(t). For the intensity of the virtual image in this case we obtain

$$I_M \sim \left| \frac{1}{\tau} \int_0^\tau A(t) dt \right|^2 = a_o^2 \left| \frac{1}{\tau} \int_0^\tau \exp[-i\delta f(t)] dt \right|^2 . \qquad (2.25)$$

This is the most general expression which describes different cases of generation for holographic fringe patterns. For instance, for the two-exposure method the time function reduces to [2.1]

$$f(t) = \begin{cases} 0 , & 0 \le t < \tau/2 \\ 1 , & \tau/2 \le t \le \tau , \end{cases} \qquad (2.26)$$

where $\tau/2$ is the duration of one exposure. Accordingly, we divide the integral (2.25) into two parts, namely

$$I_M \sim a_o^2 \left| \frac{1}{\tau} \left[\int_0^{\tau/2} dt + \int_{\tau/2}^\tau \exp(-i\delta) dt \right] \right|^2 . \qquad (2.27)$$

It can readily be verified that after integration, (2.27) reduces to

$$I_M \sim a_o^2 (1 + \cos\delta) ,$$

which coincides with (2.9) derived earlier.

The case where the phase of the object wave changes with a constant rate is of practical interest. Here, for the time function we obtain [2.26]

$$f(t) = t/\tau , \qquad (2.28)$$

where τ is the exposure duration. After substituting (2.28) into (2.25) and some algebra we arrive at

$$I_M \sim a_o^2 \left| \frac{1}{\tau} \int_0^\tau \exp\left(-i\frac{\delta}{\tau}t\right) dt \right|^2 = a_o^2 \frac{\sin^2(\delta/2)}{(\delta/2)^2} . \qquad (2.29)$$

Note that δ/τ is the rate of change of the object-wave phase at the hologram plane. The function $\sin^2(\delta/2)/(\delta/2)^2$ modulating the virtual image of the object is sketched in Fig.2.7 (curve 1). The bright-fringe intensity drops dramatically with increasing δ.

Another type of motion that can be considered is when the subject oscillates harmonically with a time function of the form

$$f(t) = \sin\omega t \qquad (2.30)$$

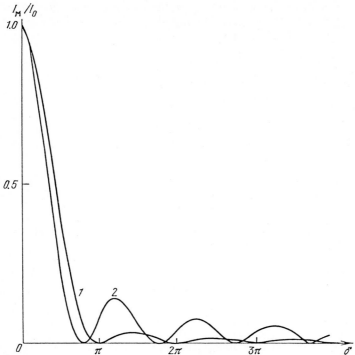

Fig.2.7. Virtual image intensity distribution for (*1*) an object moving with a constant velocity and (*2*) harmonic vibrations

where ω is the circular frequency of the oscillations. The intensity distribution of a virtual image, for a time function of the form (2.30) and $\tau \gg \omega^{-1}$, will be

$$I_M \sim a_o^2 \left| \lim_{\tau \to \infty} \frac{1}{\tau} \int_0^\tau \exp(-i\delta\sin\omega t)dt \right|^2 = a_o^2\, J_0^2(\delta)\,, \qquad (2.31)$$

where $J_0(\delta)$ is the zero-order Bessel function of the first kind. In this case the quantity δ corresponds to the vibration amplitude of the object. A plot of the $J_0^2(\delta)$ function is shown in Fig.2.7 (curve 2). *Powell* and *Stetson* [2.27,28] were the first to employ holographic interferometry for recording harmonic oscillations. The use of holographic interferometry in the investigation of vibrations will be considered in more detail in Chap.8.

To further illustrate the time-average technique, Fig.2.8a displays a holographic reconstruction of a cantilever beam loaded at a constant rate with a concentrated force, and Fig.2.8b a fringe pattern corresponding to the principal mode of vibration of the same beam. As is expected, the bright-fringe intensity drops more rapidly in the case of a constant-rate deformation.

40

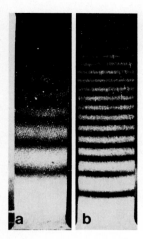

Fig.2.8a,b. Time-averaged holographic interferogram of the bending of a beam for **(a)** constant deformation rate and **(b)** harmonic vibrations

Table 2.1. Time functions and corresponding characteristic fringe functions

$f(t)$	Method	$\lvert M(\delta) \rvert^2$
0, $0 \le t \le \tau < \tau/2$ 1, $\tau/2 \le t \le \tau$	Two-exposure	$\cos^2(\delta/2)$
0, $0 \le t < \tau/m$ 1, $\tau/m \le t < 2\tau/m$ 2, $2\tau/m \le t < 3\tau/m$ .... m-1, $(m-1)\tau/m \le t \le \tau$	m-exposure	$\dfrac{\sin^2(m\delta/2)}{\sin^2(\delta/2)}$
t/τ	Time-average (motion with constant velocity)	$\dfrac{\sin^2(\delta/2)}{(\delta/2)^2}$
$\sin\omega t$	Time-average (harmonic oscillations)	$J_0^2(\delta)$

Despite the fact that (2.25) describes all the above methods of construction of holographic fringe patterns, we will use the term "time-average method" in all cases where the exposure is made of a moving or strained object. The intensity of a reconstructed image corresponding to different methods for obtaining holographic fringe patterns can be presented in a general form in terms of the so-called characteristic fringe function $M(\delta)$ [2.1]

$$I_M \sim I_0 |M(\delta)|^2 . \tag{2.32}$$

Following (2.25), the fringe function can be written as

$$M(\delta) = \frac{1}{\tau} \int_0^\tau \exp[-i\delta f(t)]dt . \tag{2.33}$$

Thus, in the holographic interferometry of diffusely reflecting objects the intensity distribution of a reconstructed image is modulated by the square of the fringe function, its form being determined by the time function. The time functions and the corresponding fringe functions, which are most frequently used in mechanics, are presented in Table 2.1. Other forms of fringe functions have been given by *Vest* [2.1].

2.2 Fringe Formation in Holographic Interferometry of Diffusely Reflecting Objects

2.2.1 Principal Relation of Holographic Interferometry

Interference fringes produced by the different methods of holographic interferometry form as a result of a change in the phase difference between interfering waves, δ, due to a deformation or displacement of the object. Therefore, the quantitative determination of these displacements of points on the surface of an object from the interference fringe patterns has practical significance. One must thus establish the relation between the fringes and the surface displacement of a strained body. We will use the so-called geometrical model [2.29-31]. In this model, a rough surface is taken to consist of mutually incoherent point scatterers. Therefore, only the waves emerging from the same points (corresponding or identical) either, a stressed or an unstressed surface will be able to interfere. Since this model presently enjoys widespread use in quantitative analysis of holographic fringe patterns, we will treat it here in some detail.

Consider the virtual images reconstructed by means of a doubly-exposed hologram as two simultaneously existing real surfaces of an object. Figure 2.9 shows the conditions of illumination and observation of the corresponding points Q and Q' on a surface element before and after its deformation. Here D is the displacement vector; S is a point light source chosen as the origin; B is the observation point; r_1, r_2, and R are the radius vectors of points Q, Q' and B, respectively; $\hat{e}_s$, $\hat{e}_1$ and $\hat{e}_2$, $\hat{e}_3$ are the unit vectors of illumination and observation for points Q and Q'. Hologram H serves to reconstruct two images of the object and also restricts the cone of possible directions from which the surface element of interest can be observed. In place of the phase difference between light waves at the hologram plane we will now assume the phase shift δ to be due to the point Q on the surface having been displaced in the given direction of observation.

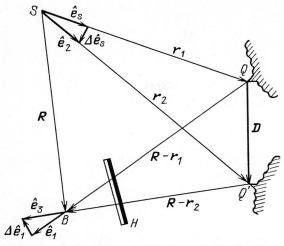

Fig.2.9. Interferometer arrangement for derivation of the principal relation of holographic interferometry

The path difference Δ between the rays from the light source S passing through the corresponding points Q and Q′ to the observation point B can be written as

$$\Delta = SQ'B - SQB . \tag{2.34}$$

Using the nomenclature of Fig.2.9 we can write the optical paths in the following way

$$SQB = \hat{e}_s \cdot r_1 + \hat{e}_1 \cdot (R - r_1) , \tag{2.35}$$

$$SQ'B = \hat{e}_2 \cdot r_2 + \hat{e}_3 \cdot (R - r_2) . \tag{2.36}$$

Substituting (2.35,36) into (2.34), we obtain an expression for the path difference

$$\Delta = \hat{e}_2 \cdot r_2 + \hat{e}_3 \cdot (R - r_2) - \hat{e}_s \cdot r_1 - \hat{e}_1 \cdot (R - r_1) . \tag{2.37}$$

The unit vectors of illumination and observation for the point Q′ can be represented as

$$\hat{e}_2 = \hat{e}_s + \Delta\hat{e}_3, \quad \hat{e}_3 = \hat{e}_1 + \Delta\hat{e}_1 . \tag{2.38}$$

substituting (2.38) into (2.37), we obtain after some algebra:

$$\Delta = (\hat{e}_1 - \hat{e}_s) \cdot (\hat{r}_2 - \hat{r}_1) + \Delta\hat{e}_s \cdot \hat{r}_2 + \Delta\hat{e}_1 \cdot (R - r_2) . \tag{2.39}$$

In optical systems of the holographic interferometers the distance from the light source to the object is, as a rule, much greater than the displacement $|D|$

$$|D| = |r_2 - r_1| \ll |r_2| \simeq |r_1| \;,$$

and, therefore, for the vectors we may safely assume

$$\Delta\hat{e}_s \perp r_2 \;, \quad \Delta\hat{e}_1 \perp (R - r_2) \;.$$

With this assumption, expression (2.39) acquires the form

$$\Delta = (\hat{e}_1 - \hat{e}_s) \cdot D \;. \tag{2.40}$$

The phase difference of the interfering rays, δ, is related to the optical path difference through the wave number $2\pi/\lambda$

$$\delta = (2\pi/\lambda)(\hat{e}_1 - \hat{e}_s) \cdot D \;, \tag{2.41}$$

where λ is the wavelength of light. Bright interference fringes are observed for

$$\delta = 2\pi n \quad (n = 0, 1, 2, \ldots) \tag{2.42}$$

and dark ones

$$\delta = 2\pi(n - \tfrac{1}{2}) \quad (n = 1, 2, 3, \ldots) \;, \tag{2.43}$$

where n is the absolute order of a bright (dark) fringe at the surface point of interest.

Expression (2.41) can be rewritten in the following way for bright fringes:

$$n\lambda = (\hat{e}_1 - \hat{e}_s) \cdot D \quad (n = 0, 1, 2, \ldots) \;, \tag{2.44}$$

and for the dark ones:

$$(n - \tfrac{1}{2})\lambda = (\hat{e}_1 - \hat{e}_s) \cdot D \quad (n = 1, 2, 3, \ldots) \;. \tag{2.45}$$

Equation (2.44), or (2.45), which establishes the relation between the displacement vector of a surface point, the parameters of the interferometer system (directions of illumination and observation), and the absolute order of the fringe at this point on the pattern, is frequently called the *principal relation of holographic interferometry*. For the sake of uniqueness, we will consider in the following discussion only the equation for bright fringes (2.44) bearing in mind that all the subsequent reasoning and conclusions will also be valid for the dark fringes described by (2.45).

We introduce the concept of the *sensitivity vector* **k** as the difference between the unit vectors for observation and illumination [2.29, 30]

$$k = \hat{e}_1 - \hat{e}_s \;. \tag{2.46}$$

44

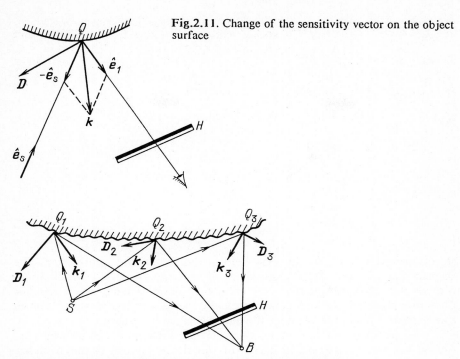

Fig.2.11. Change of the sensitivity vector on the object surface

Now (2.44) can be converted to a form more convenient for the analysis

$$n\lambda = \mathbf{k} \cdot \mathbf{D} . \qquad (2.47)$$

The physical meaning of (2.47) can be derived by considering the scheme of illumination and observation of a surface point Q shown in Fig.2.10. The direction of illumination of the point Q is given by the unit vector $\hat{\mathbf{e}}_s$, and that of observation through the hologram H by the unit vector $\hat{\mathbf{e}}_1$. We assume that the displacement vector D of the point of interest lies in the plane defined by the unit vectors $\hat{\mathbf{e}}_s$ and $\hat{\mathbf{e}}_1$. As seen in this configuration, the sensitivity vector is directed along the bisector of the angle between the directions of illumination and observation. Thus, the fringe pattern seen from a fixed observation point permits determination of the projection of the displacement vector onto the sensitivity vector by means of (2.47). This implies that in the general case of object deformation the interference fringes do not lend themselves to a direct mechanical interpretation. This introduces serious difficulties into a quantitative analysis of holographic fringe patterns due to diffusely reflecting bodies. The pattern proper of the fringes is affected by the variation over the surface of both magnitude and direction of the sensitivity vector and those of the displacement vector. This is clearly seen from the scheme in Fig.2.11. For given surface points Q_1, Q_2, Q_3 illuminated by a point source S and observed from a point B, the

45

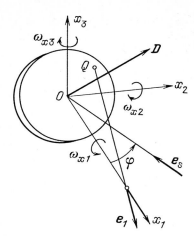

Fig.2.12. Interferometer arrangement for studying fringe patterns originating from translations and rotations of a disc

sensitivity vectors $\mathbf{k}_1$, $\mathbf{k}_2$, $\mathbf{k}_3$ vary both in magnitude and in direction. The displacement vectors $\mathbf{D}_1$, $\mathbf{D}_2$ and $\mathbf{D}_3$ at these points likewise have different directions and magnitude. By using collimated illumination and a telecentric observation system, however, the constancy in the sensitivity vector both in direction and in magnitude over the entire surface is ensured.

The principal relation of holographic interferometry can also be employed for the analytical calculation of the fringe pattern for known displacement-vector fields and parameters of the optical system of the holographic interferometer used. By representing the displacement and sensitivity vector components in Cartesian coordinates (x_1, x_2, x_3) attached to the object, one can write (2.47) in the following way

$$k_{x1} D_{x1} + k_{x2} D_{x2} + k_{x3} D_{x3} = n\lambda \,, \tag{2.48}$$

where k_{x1}, k_{x2}, k_{x3} are the components of the sensitivity vector, and D_{x1}, D_{x2}, D_{x3} are those of the displacement vector of the surface point in question. The quantity that one must derive from (2.48) is the fringe order at the surface point.

Consider now the fringe patterns which will be generated by the simplest kind of object displacement, namely, a translation and a rotation. This is not only of purely methodological interest; the knowledge of the actual pattern to be expected in a rigid displacement of an object also has practical importance. Figure 2.12 shows a simple disc-shaped object. We choose a Cartesian system (x_1, x_2, x_3) such that the x_1 axis is normal to the disc surface, and the x_2 and x_3 axes lie in the plane of its surface. The disc surface is illuminated by a plane wave, its unit vector $\hat{\mathbf{e}}_s$ being parallel to the x_1, x_2 plane and making an angle ϕ with the x_1 axis. The observation point B lies on the x_1 axis at a distance R from the origin. This holographic interferometer arrangement is frequently used in practice.

The components of the unit observation $(\hat{\mathbf{e}}_1)$ and illumination $(\hat{\mathbf{e}}_s)$ vectors for an arbitrary surface point $Q(0, x_2, x_3)$ will be described by the following expressions:

46

$$e^1_{x1} = R(R^2 + x_2{}^2 + x_3{}^2)^{-1/2} , \quad e^s_{x1} = - \cos\varphi ,$$

$$e^1_{x2} = x_2(R^2 + x_2{}^2 + x_3{}^2)^{-1/2} , \quad e^s_{x2} = - \sin\varphi ,$$

$$e^1_{x3} = x_3(R^2 + x_2{}^2 + x_3{}^2)^{-1/2} , \quad e^3_{x3} = 0 . \tag{2.49}$$

Using (2.49, 46) one can readily determine the sensitivity vector components k_{x1}, k_{x2}, k_{x3} for this point:

$$k_{x1} = R(R^2 + x_2{}^2 + x_3{}^2)^{-1/2} + \cos\varphi , \tag{2.50}$$

$$k_{x2} = x_2(R^2 + x_2{}^2 + x_3{}^2)^{-1/2} + \sin\varphi , \tag{2.51}$$

$$k_{x3} = x_3(R^2 + x_2{}^2 + x_3{}^2)^{-1/2} . \tag{2.52}$$

With the disc translated along the x_2 axis, the displacement vectors of all points on the surface will be the same:

$$\mathbf{D} = (0, D_{x2}, 0) .$$

Under these conditions only the second term in (2.48) is nonzero. Using (2.51), we obtain

$$[x_2(R^2 + x_2{}^2 + x_3{}^2)^{-1/2} + \sin\varphi]D_{x2} = n\lambda . \tag{2.53}$$

Bearing in mind that usually x_2, and $x_3 \ll R$, (2.53) can be rewritten as

$$(x_2/R + \sin\varphi)D_{x2} = n\lambda . \tag{2.54}$$

Thus the interference fringes generated by an object translated in its plane in the direction of the x_2 axis represent equidistant straight lines parallel to the x_3 axis. For the spacing between the fringes $L(D_{x2})$ we obtain

$$L(D_{x2}) = (\partial n/\partial x_2)^{-1} = \lambda R/D_{x2} . \tag{2.55}$$

Figure 2.14c displays a typical fringe pattern corresponding to this case.

For translation along the x_3 axis, $\mathbf{D} = (0, 0, D_{x3})$, and the only nonzero term in the fringe equation (2.48) is the one with the sensitivity-vector component k_{x3} given by (2.52), namely

$$x_3 D_{x3}/R = n\lambda . \tag{2.56}$$

This equation describes equidistant straight lines parallel to the x_2 axis. The expression for the fringe spaceing $L(D_{x3})$ is similar to (2.55):

$$L(D_{x3}) = \lambda R/D_{x3} . \tag{2.57}$$

47

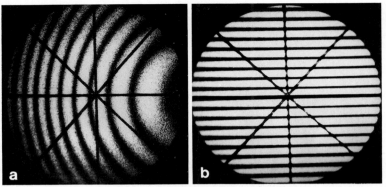

Fig.2.13. Interference fringes obtained by (a) translating a disc perpendicular to its surface, and (b) rotating it about x_2 axis

For a translation normal to the surface of the object, i.e. $D = (D_{x1},0,0)$, (2.48) yields an erroneous result. To obtain an equation describing the experimentally observed fringe pattern, one must use the exact expression for the path difference, (2.34), i.e. take into account second-order terms omitted in the derivation of (2.40). In this case the fringe equation will be [2.32]

$$A(x_2^2+x_3^2) + Bx_2 + C = 0 , \qquad (2.58)$$

where A, B and C are defined in terms of the parameters of the interferometric system employed. Thus, in a translation normal to the surface of an object, the interference fringes will be concentric circles with the center shifted along the x_2 axis by $-B/2A$. A typical fringe pattern corresponding to this type of displacement is shown in Fig.2.13a.

The displacement vector of surface points on the disc rotated about the x_2 axis by a small angle ω_{x2} is $D = (\omega_{x2}x_3,0,0)$. For the fringe equation in this case we obtain

$$x_3(1 + \cos\varphi)\omega_{x2} = n\lambda . \qquad (2.59)$$

When the object is rotated about an axis lying in its surface plane the fringes will be equidistant straight lines parallel to the rotation axis. The fringe spacing can be determined from the expression

$$L(\omega_{x2}) = \lambda[(1 + \cos\varphi)\omega_{x2}]^{-1} . \qquad (2.60)$$

A typical fringe pattern for a disc rotated about the x_2 axis is shown in Fig.2.13b. One can readily see that the fringes created by the object rotated about the x_3 axis are parallel to this axis.

Rotation of the disc about the x_1 axis by a small angle ω_{x1} is described by the following displacement vector of the surface points:

$$\mathbf{D} = (0,-\omega_{x1}x_3,\omega_{x1}x_2) . \qquad (2.61)$$

48

Substituting (2.51,52 and 61) into (2.48) we obtain after some manipulation

$$\omega_{x1} x_3 \sin\varphi = n\lambda . \tag{2.62}$$

Equation (2.62) describes equidistant straight fringes parallel to the x_2 axis; the fringe spacing $L(\omega_{x1})$ can be found from

$$L(\omega_{x1}) = \lambda(\omega_{x1} \sin\varphi)^{-1} . \tag{2.63}$$

A remarkable feature of the interference fringes associated with this kind of rotation is their high sensitivity to a change in the direction of observation. As the observation direction changes within the hologram aperture, both direction and spacing of the fringes change [2.33, 34].

2.2.2 Fringe Localization and Visibility

Interference fringes formed in displacement and deformation of an object with a diffusely reflecting surface are *localized* in the general case. When the light waves recorded, for instance, on a doubly-exposed hologram are reconstructed, the observer can see interference fringes behind or in front of the surface image, and only in some particular cases on the surface proper. This complex phenomenon has been the subject of many studies, among them are [2.2, 29, 35–38]. On the basis of the concepts of ray optics, in this subsection we are going to dwell on the major ideas for fringe localization, essential for solving problems in mechanics.

The phenomenon of fringe localization is illustrated in Fig.2.14 showing reconstructed doubly-exposed holograms of a virtual image of the same circular disc recorded under different conditions. The fringe pattern was obtained with the configuration presented in Fig.2.12 with the disc translated along the x_2 axis. When the camera is focussed on the disc surface with the lens set at f/2.8, no interference fringes are observed (Fig.2.14a). For convenience, in focussing the image, a system of lines passing through the center is drawn on the surface. A sharp image of the fringes is obtained at a distance of about 2 m behind the object. (Fig.2.14b). If we now focus the camera on the disc surface and reduce the lens aperture to f/16, high-contrast fringes will be observed together with a sharp image of the object surface (Fig.2.14c). This experiment, which is frequently described in the literature, graphically illustrates the substantial effect that the aperture of the observation system has on the fringe localization.

Fringe localization can be also observed in the reconstruction of a real image. When the entire surface of the same doubly-exposed hologram is illuminated by a reference conjugate, no interference fringes will be seen on the disc surface. However, there exists a region of high fringe contrast some distance away from the disc image. If we now illuminate the hologram with a nonexpanded laser beam, fringes will appear on the reconstructed real image. In this particular case the illuminated area of the hologram surface plays the role of the observation aperture.

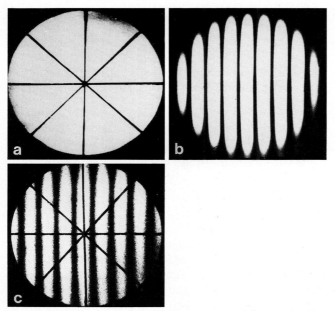

Fig.2.14a-c. An image of a disc reconstructed from a doubly exposed hologram under different photographing conditions: (**a**) camera focussed on the disc surface; (**b**) camera focussed on the localization plane; (**c**) small aperture camera focussed on the disc surface

The mechanism of fringe localization can be easily understood when one examines the object image reconstructed with a doubly-exposed hologram in the configuration shown in Fig.2.15. Hologram H, illuminated with a reference wave from source S_0, reconstructs two object waves reflected by the surface in the original and deformed states. The objective lens F is focused onto a point B in front of the object image and generates its image B' on screen M. In accordance with the geometric model considered in Sect.2.2.1, the fringe pattern on the object's image is formed by interference of rays from the corresponding points. Point B will receive light rays from a surface area dS, its size being determined by the camera lens aperture. For an arbitrary point B the phase difference of rays coming from the corresponding points on the surface element dS will be different for each pair of points. As a result, in the vicinity of this point the fringes will have low contrast or none at all. However, for a chosen observation direction there will exist a point B where rays from all correspond points of the element dS will arrive with approximately the same phase difference. It is in the vicinity of such a point that interference fringes will appear. To determine its position, one has to locate, along the chosen observation direction, the point where the variations of the phase difference from the corresponding points of the surface element are zero. Assuming a plane surface of the object and choosing the Cartesian system x_1, x_2, x_3 such that the x_1 axis will be normal to the surface, and x_2 and x_3 are on the surface, the condition for localization can be written as [2.1].

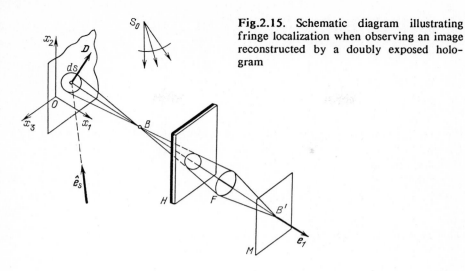

Fig.2.15. Schematic diagram illustrating fringe localization when observing an image reconstructed by a doubly exposed hologram

$$d\delta = \frac{\partial\delta}{\partial x_2}dx_2 + \frac{\partial\delta}{\partial x_3}dx_3 = 0 \ . \tag{2.64}$$

In the case when the observation aperture is approximately of the same size in all directions, e.g. circular, condition (2.64) can be satisfied by independently varying dx_2 and dx_3 resulting in the following localization condition:

$$\frac{\partial\delta}{\partial x_2} = 0 \ , \quad \frac{\partial\delta}{\partial x_3} = 0 \ . \tag{2.65}$$

Another approach to the determination of the localization position based on an analysis of the object-diffracted field and which directly takes into account the aperture of the system, was developed by *Stetson* [2.39]. The position of fringe localization is determined from the condition

$$\frac{\partial\delta}{\partial e_{x2}^1} = 0 \ , \quad \frac{\partial\delta}{\partial e_{x3}^1} = 0 \ , \tag{2.66}$$

where e_{x2}^1, e_{x3}^1 are components of the unit observation vector. For collimated illumination, the criteria (2.65,66) yield the same fringe localization conditions.

In our derivation of the fringe-localization equations we will use condition (2.65) and follow the approach used by *Vest* [2.1]. We write the components of the vectors $\hat{e}_1$, $\hat{e}_S$, D and k (Fig.2.15) for an arbitrary point on the object surface $Q(O,x_2,x_3)$ in the coordinate system chosen as

$$\hat{e}_1 = (e_{x1}^1, e_{x2}^1, e_{x3}^1) \, ,$$

$$\hat{e}_S = (e_{x1}^S, e_{x2}^S, e_{x3}^S) \, ,$$

$$\mathbf{D} = (D_{x1}, D_{x2}, D_{x3}) \, ,$$

$$\mathbf{k} = (k_{x1}, k_{x2}, k_{x3}) \, .$$

(2.67)

We can rewrite the principal relation of holographic interferometry (2.41) for the point in question in terms of the displacement- and sensitivity-vector components.

$$\delta = \frac{2\pi}{\lambda}(k_{x1}D_{x1} + k_{x2}D_{x2} + k_{x3}D_{x3}) \, .$$

(2.68)

The localized equation (2.65) can now be represented in the form:

$$\frac{\partial k_{x1}}{\partial x_2}D_{x1} + \frac{\partial k_{x2}}{\partial x_2}D_{x2} + \frac{\partial k_{x3}}{\partial x_2}D_{x3} + k_{x1}\frac{\partial D_{x1}}{\partial x_2} + k_{x2}\frac{\partial D_{x2}}{\partial x_2} + k_{x3}\frac{\partial D_{x3}}{\partial x_2} = 0 \, ,$$

(2.69)

$$\frac{\partial k_{x1}}{\partial x_3}D_{x1} + \frac{\partial k_{x2}}{\partial x_3}D_{x2} + \frac{\partial k_{x3}}{\partial x_3}D_{x3} + k_{x1}\frac{\partial D_{x1}}{\partial x_3} + k_{x2}\frac{\partial D_{x2}}{\partial x_3} + k_{x3}\frac{\partial D_{x3}}{\partial x_3} = 0 \, .$$

(2.70)

Limiting ourselves, for instance, to the case of an object surface illuminated by a plane wave, i.e., when the vector $\hat{e}_S$ is constant over the entire surface, we obtain

$$\frac{\partial k_{xi}}{\partial x_j} = \frac{\partial e_{xi}^1}{\partial x_j}$$

(2.71)

where i = 1,2,3; j = 2,3. Expressing the observation-vector components in terms of the coordinates of the surface point in question, Q, and of the observation point, $B(x_{10}, x_{20}, x_{30})$, we will have

$$e_{x1}^1 = x_1[x_1^2 + (x_{20}-x_2)^2 + (x_{30}-x_3)^2]^{-1/2} \, ,$$

$$e_{x2}^1 = (x_{20}-x_2)[x_1^2 + (x_{20}-x_2)^2 + (x_{30}-x_3)^2]^{-1/2} \, ,$$

(2.72)

$$e_{x3}^1 = (x_{30}-x_3)[x_1^2 + (x_{20}-x_2)^2 + (x_{30}-x_3)^2]^{-1/2} \, .$$

Next we use (2.72) to define the derivatives of the sensitivity-vector components (2.71)

$$\frac{\partial k_{x1}}{\partial x_2} = (e_{x1}^1)^2 e_{x2}^1 / x_1 \ ,$$

$$\frac{\partial k_{x1}}{\partial x_3} = (e_{x1}^1)^2 e_{x3}^1 / x_1 \ ,$$

$$\frac{\partial k_{x2}}{\partial x_2} = - e_{x1}^1 [(e_{x1}^1)^2 + (e_{x3}^1)^2] / x_1 \ ,$$

$$\frac{\partial k_{x2}}{\partial x_3} = e_{x1}^1 e_{x2}^1 e_{x3}^1 / x_1 \ ,$$

$$\frac{\partial k_{x3}}{\partial x_2} = e_{x1}^1 e_{x2}^1 e_{x3}^1 / x_1 \ ,$$

$$\frac{\partial k_{x3}}{\partial x_3} = - e_{x1}^1 [(e_{x1}^1)^2 + (e_{x2}^1)^2] / x_1 \ .$$

(2.73)

Substituting (2.73) into (2.69, 70), we obtain after some transformations, two fringe localization conditions:

$$x_1 = \frac{e_{x1}^1 \{ e_{x1}^1 e_{x2}^1 D_{x1} - [(e_{x1}^1)^2 + (e_{x3}^1)^2] D_{x2} + e_{x2}^1 e_{x3}^1 D_{x3} \}}{k_{x1} \partial D_{x1} / \partial x_2 + k_{x2} \partial D_{x2} / \partial x_2 + k_{x3} \partial D_{x3} / \partial x_2} \ , \qquad (2.74)$$

$$x_2 = \frac{e_{x1}^1 \{ e_{x1}^1 e_{x3}^1 D_{x1} + e_{x2}^1 e_{x3}^1 D_{x2} - [(e_{x1}^1)^2 + (e_{x2}^1)^2] D_{x3} \}}{k_{x1} \partial D_{x1} / \partial x_3 + k_{x2} \partial D_{x2} / \partial x_3 + k_{x3} \partial D_{x3} / \partial x_3} \ . \qquad (2.75)$$

Both (2.74 and 75) describe a surface in space. The line which intersects both these surfaces satisfies both equations. Thus, when viewed through a circular aperture, the interference fringes in holographic interferometry of diffusely reflecting objects are, in general, localized along a line in space.

We now calculate the intensity at point B_1 created by all pairs of corresponding points on the surface element dS (Fig. 2.15). For a rectangular aperture and a corresponding rectangular surface element dS with sides $2\Delta x_2$ and $2\Delta x_3$, we can use (2.9) describing fringes in the two-exposure method to calculate the intensity $I(B_1)$:

$$I(B_1) = \int_{x_2^q - \Delta x_2}^{x_2^q + \Delta x_2} \int_{x_3^q - \Delta x_3}^{x_3^q + \Delta x_3} [1 + \cos\delta(x_2, x_3)] dx_2 \, dx_3 \ , \qquad (2.76)$$

where $(0, x_2^q, x_3^q)$ are the coordinates of the point Q at the center of the element dS. Replacing $\delta(x_2, x_3)$ with the first term of the Taylor expansion in the vicinity of the point and calculating the integral (2.76), we obtain [2.1]

$$I(B_1) = 4\Delta x_2 \Delta x_3 \left\{ 1 + \frac{\sin[\partial\delta(x_2^q, x_3^q)/\partial x_2]\Delta x_2}{[\partial\delta(x_2^q, x_3^q)/\partial x_2]\Delta x_2} \right.$$

$$\left. \times \frac{\sin[\partial\delta(x_2^q, x_3^q)/\partial x_3]\Delta x_3}{[\partial\delta(x_2^q x_3^q)/\partial x_3]\Delta x_3} \cos\delta(x_2^q, x_3^q) \right\} . \qquad (2.77)$$

The visibility V of the fringes described by (2.77) in the vicinity of the point B_1 is defined by

$$V = \frac{I_{max} - I_{min}}{I_{max} + I_{min}} = \left| \frac{\sin[\partial\delta(x_2^q, x_3^q)/\partial x_2]\Delta x_2}{[\partial\delta(x_2^q, x_3^q)/\partial x_2]\Delta x_2} \frac{\sin[\partial\delta(x_2^q x_3^q)/\partial x_3]\Delta x_3}{[\partial\delta(x_2^q, x_3^q)/\partial x_3]\Delta x_3} \right| . \qquad (2.78)$$

As seen from (2.78), the fringes will have maximum contrast for

$$\partial\delta(x_2^q, x_3^q)/\partial x_2 = 0 , \quad \partial\delta(x_2^q, x_3^q)/\partial x_3 = 0 . \qquad (2.79)$$

The conditions of maximum contrast, (2.79), coincide with the criterion of fringe localization (2.65). Hence, when viewed in a fixed direction, the interface fringes are localized in the regions where their visibility is maximum.

The extent of the fringe-localization region depends on the size of the viewing aperture. As it decreases, the extent of the localization region grows, with the fringes appearing at the surface of the object. This has a particular, practical significance for measurements of surface displacements of deformed objects, since the methods of interpretation of holographic fringe patterns presently accepted are based on an analysis of the fringes on the surface of the object. In doing this one should bear in mind, however, that as the aperture is reduced, the speckle in an image reconstructed with coherent light increases in size with a corresponding degradation in fringe resolution. This precludes the use of aperture reduction in cases where the fringes in the localization region have a higher frequency than, for instance, in the areas of strain concentration.

A method of tying localized fringes to the object surface based on the use of the so-called observer-projection theorem has been developed by *Stetson* [2.25]. The theorem, presented here without proof, reads as follows: the fringes localized on the object surface can be projected onto it radially from the center of the viewing aperture. This theorem implies that one can photograph the fringes in the localization region with a large aperture camera. Next, without changing the camera's position, the object surface is photographed. After this a magnifier can be used to match the fringe pattern with the object surface. The fringe pattern thus obtained will be equivalent to the one that would be observed on the object surface with a small aperture. Figure 2.16 displays a scheme illustrating the concept of radial projection of localized interference fringes onto the object surface. In this case speckle affects the fringe resolution only to a limited extent.

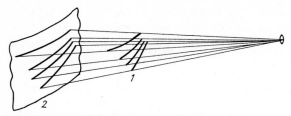

Fig.2.16. Projecting localized interference fringes 1 on the object surface 2

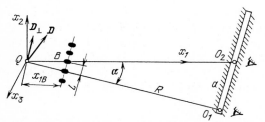

Fig.2.17. Localization and parallax of interference fringes with a slit aperture

When viewing a reconstructed image through a slit aperture, the quantities dx_2 and dx_3 in (2.64) will no longer be independent. In this case, the interference fringes will be localized on a surface. Figure 2.17 illustrates the arrangement where the fringes are viewed through a slit aperture oriented normal to the x_1 axis. *Stetson* [2.25] derived an important relation for this fringe-observation arrangement:

$$D_\perp = \lambda x_{1B}/L \, , \tag{2.80}$$

where $D_\perp$ is the displacement vector component normal to the viewing direction and parallel to the slit; x_{1B} is the distance from the point Q on the object surface to the fringe localization surface; L is the fringe spacing in the direction parallel to the slit on the localization surface where it is intersected by the x_1 axis. Thus, according to (2.80), the fringe localization position and the fringe spacing permit one to determine the displacement vector component parallel to the slit aperture and normal to the viewing direction.

For a better understanding of the main techniques used in the interpretation of holographic-fringe patterns, we consider the fringe parallax with the arrangement of Fig.2.17. When the viewing position changes from a point O_1 of the slit aperture to a point O_2, the observer sees only one fringe pass through the surface point. In the general case, different numbers of fringes may pass through the point Q depending on the distance a. For small scanning angles α the geometry of Fig.2.17 suggests an equivalence in localization and parallax of fringes, first derived by *Stetson* [2.40]:

$$x_{1B}/L \simeq R/a \, ,$$

where R and a are defined in Fig.2.17. Their ratio is a measure of the displacement normal to the bisector of the angle between the viewing directions passing through the points O_1 and O_2 [2.31]. In other words, fringe parallax is observed only in the cases where the fringes are not localized on the surface of the object.

Consider now fringe localization for the simplest kinds of translation and rotation of an object. When the object is illuminated by a plane wave and the reconstructed image is viewed with a telecentric system, the fringes due to a translation $\mathbf{D}_1 = (D_{x1},0,0)$, $\mathbf{D}_2 = (0,D_{x2},0)$, and $\mathbf{D}_3 = (0,0,D_{x3})$ will be localized at infinity, as this can be readily seen from (2.74,75). If the object is illuminated by a spherical wave, translation creates fringes localized at a finite distance from the object surface [2.33,34]. This case of localization for translation in the object's surface plane is shown in Fig.2.14.

When an object is rotated about an axis lying in the surface plane, x_3 (Fig.2.12), the displacement vector can be written as $\mathbf{D} = (\omega_{x3}x_2,0,0)$. Under the condition of collimated illumination of a disc and viewing with a telecentric system, (2.74,75) can be written as [2.1]

$$x_1 = (e_{x1}^1)^2 e_{x2}^1 x_2/k_{x1} \; , \tag{2.81}$$

$$0 = (e_{x1}^1)^2 e_{x3}^1 \omega_{x3} x_2 \; . \tag{2.82}$$

The components of the unit observation vector are then given by $\hat{e}_1 = (1,0,0)$, Eq.(2.82) becomes an identity, and (2.81) yields the fringes localized on the disc surface.

When the object is rotated about the x_1 axis normal to the object surface by an angle ω_{x1} (Fig.2.12), the displacement vector of the surface points is $\mathbf{D} = (0,-\omega_{x1}x_3,\omega_{x1}x_2)$. For the fringe localization equations (2.74, 75), corresponding to this displacement we will have [2.1]

$$x_1 = e_{x1}^1\{[(e_{x1}^1)^2 + (e_{x3}^1)^2]x_3 + e_{x2}^1 e_{x3}^1 x_2\}/k_{x3} \; , \tag{2.83}$$

$$x_1 = e_{x1}^1\{e_{x2}^1 e_{x3}^1 x_3 + [(e_{x1}^1)^2 + (e_{x2}^1)^2]x_2\}/k_{x2} \; . \tag{2.84}$$

For the illumination and observation conditions assumed here, $\hat{e}_S = (-\cos\varphi, -\sin\varphi,0)$ and $\hat{e}_1 = (1,0,0)$, the equations reduce to the form

$$x_3 = 0 \; , \quad x_1 = x_2/\sin\varphi \; . \tag{2.85}$$

The interference fringes corresponding to this kind of rotation are localized on the line $x_1 = (\sin\varphi)^{-1}x_2$ in the $x_3 = 0$ plane.

Figure 2.18a presents a fringe on a disc surface observed through a large-aperture optical system. The disc was rotated, between two successive exposures, through a small angle about an axis normal to the surface. The fringes are seen in a small area at the disc centre. If we now focus the camera onto a plane behind the disc surface, the fringe-localization region

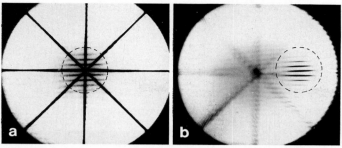

Fig.2.18a,b. Fringe localization with the disc rotated about a surface normal when the camera is focussed on (**a**) the disc surface, and (**b**) on a plane behind the disc surface. The dashed circles identify the fringe position

shifts (Fig.2.18b). This implies that the fringes are localized on a straight line, in accordance with the theoretical relation (2.85).

2.2.3 Strain-Induced Fringes

An analysis of stressed states in solid mechanics involves consideration of individual components of the displacement vector, namely, those normal and tangential to the object surface. One can classify deformations according to which one of these components prevails. Normal displacement (deflection) is predominant in the bending of plates and beams, and tangential displacement in the plane deformation of objects.

The smallness of the displacements measurable by the methods of holographic interferometry allows us to use the relations derived in the theory of elasticity for the treatment of experimental data (Chap.5). Combined application of those solutions and of the principal equations of fringe formation yields theoretical relations describing the shape of the interference fringes and their localizations.

Figure 2.19 illustrates the application of a tensile stress by a force P to a cantilever beam of rectangular cross section. The coordinate system is attached to the side surface of the beam, its origin being at the point of fixing. The x_2 axis is directed along the beam axis, and x_1 and x_3 are in the plane perpendicular to P. The holograms were recorded with the beam illuminated by a plane wave and the fringes are viewed with a telecentric

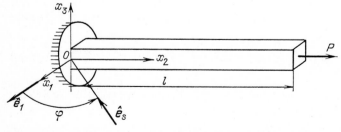

Fig.2.19. Interferometer setup to study the fringe pattern for a cantilever beam under tension

system. The unit illumination ($\hat{e}_S$) and observation ($\hat{e}_1$) vectors in the chosen coordinate will be

$$\hat{e}_S = (-\cos\varphi, -\sin\varphi, 0) , \tag{2.86}$$

$$\hat{e}_1 = (1, 0, 0) . \tag{2.87}$$

Using (2.86, 87) one can write for the sensitivity vector components

$$\mathbf{k} = \hat{e}_1 - \hat{e}_S = (1+\cos\varphi, \sin\varphi, 0) . \tag{2.88}$$

The displacement of points on the beam surface along the axis then is

$$\mathbf{d} = (0, bx_2, 0) , \tag{2.89}$$

where $b = P/EF$, and EF is the longitudinal rigidity.

The fringe equation (2.48) for the displacement vector (2.89) and sensitivity vector (2.88) will acquire the following form:

$$bx_2 \sin\varphi = n\lambda . \tag{2.90}$$

Equation (2.90) describes a system of equidistant fringes normal to the beam axis. The fringe spacing L can be determined from

$$L = (\partial n/\partial x_2)^{-1} = \lambda/b\sin\varphi . \tag{2.91}$$

For a beam under tensile stress, (2.74, 75) yield the fringe localization condition:

$$x_1 = - (\sin\varphi)^{-1} x_2 , \quad x_3 = 0 . \tag{2.92, 93}$$

As follows from these expressions, the fringes are localized in the $x_3 = 0$ plane along the line defined by (2.92). Obviously, as the viewing direction varies within the hologram aperture cone, the fringes will shift relative to the cantilever beam image, in other words, fringe parallax will be observed.

When a force P is applied in the x_1, x_2 plane to the tip of a cantilever beam of length ℓ (Fig.2.19), the deflection of the elastic axis of the beam is described by the well known cubic relation. Since the deflections of the beam surface and of its axis practically coincide, one can write for the displacement vector:

$$\mathbf{d} = [a(3x_2^2 - x_2^3/\ell), 0, 0] , \tag{2.94}$$

where $a = P\ell/6EJ$; EJ is the beam flexural rigidity. Substituting (2.94 and 88) into (2.48) yields the fringe equation:

$$a(3x_2^2 - x_2^3/\ell)(1 + \cos\varphi) = n\lambda . \tag{2.95}$$

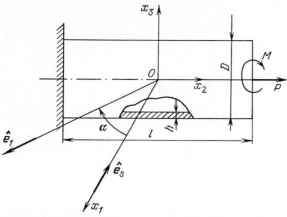

Fig.2.20. Interferometer arrangement to study the fringe pattern caused by deformation of a cylindrical shell

This relation describes a fringe pattern corresponding to the holographic interferogram shown in Fig.2.1a.

Equations (2.74,75) yield the relation $x_1 = 0$ for the case of beam deflection, implying that the fringes localize on the image of the beam surface. Indeed, as the viewing direction is varied within the hologram aperture, no fringe parallax is observed in the pattern on the beam surface. The above approach was also used in fringe-pattern calculations for objects of a more complex shape, in particular for cylindrical shells [2.41, 42].

An a-priori idea of the expected fringe pattern can be very helpful in planning an experiment and designing the object loading conditions since it facilitates an analysis of the results obtained. Figure 2.20 presents a holographic interferometer arrangement used in calculating the fringe pattern on the surface of a short cylindrical shell of diameter D = 60 mm, length ℓ = 100 mm and wall thickness h = 1.5 mm made of D16T alloy. The loading device should be capable of subjecting the shell to axial tension and torsion about its axis. Figure 2.21a depicts a fringe pattern ($\alpha = 20°$) calculated for the shell under a tensile stress of P = 3.2 kN and plotted along the involute of the shell surface. The radial and axial components of the displacement vector for the surface points were determined by the initial-parameter method [2.43]. For comparison, Fig.2.21b shows an experimental fringe pattern obtained in the two-exposure arrangement of Denisyuk. The calculated and experimental fringe patterns obtained for the same shell under torsion of M = 8 Nm are depicted in Figs.2.21c and d, respectively. A qualitative comparison of the calculated and experimental data reveal good agreement for both loading regimes, which implies that the required loading conditions for the cylidrical shall have been met. Equations (2.74,75) cannot be used to evaluate the fringe localization in this case since they were obtained for flat objects illuminated with collimated light and viewed with a telecentric system. For cylindrical shells, fringe localization can be studied by means of the equations derived by *Prikryl* [2.35].

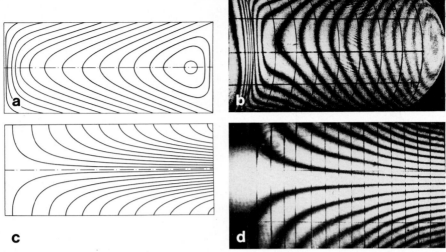

Fig.2.21a-d. Calculated and experimental fringe patterns for a shell subjected to (a,b) a tensile stress, and (c,d) a bending moment

2.2.4 Fringe Compensation for Rigid Displacements

Deformation of a structural element can be accompanied by an overall rigid-body displacement. The fringe pattern then contains not only information on the localized deformation but also on the total displacement of the object. This represents the major difficulty in holographic interferometry for measurements of the displacements of points on the surface of a stressed object. In the following discussion we will discriminate between *strain-induced* and *rigid-body* displacements such as translation, rotation and their combinations. In the general case, a holographic interferogram contains information on the total displacement vector

$$\mathbf{D} = \mathbf{D_0} + \mathbf{d} \, , \qquad (2.96)$$

where **d** is the strain-induced displacement vector, and $\mathbf{D_0}$ is the rigid-body displacement vector. It should be noted that rigid-body displacement of an object can also affect the fringe localization. Therefore the possibility of isolating strain-induced displacements against a background of rigid-body motion determines whether holographic interferometry can be employed in a quantitative study of the deformation. We now consider the most common techniques which permit restriction or complete exclusion of the effect of overall motions of an object on the detection of strain-induced displacement.

One of the main techniques is simply to design a loading device which ensures the required deformation of an object while preventing its overall displacement. It is crucial to the fine adjustment of the loading device that one can measure the displacements of its various elements (columns, holders, rods, etc.) through holographic tests. The shape of the corresponding fringes, their localization, and behavior under variation of the viewing direction within the aperture subtended by the hologram permit not only qualitative but also, in some cases, even quantitative evaluation of the displacements and rotations of these structural elements. For this purpose, one can utilize the results presented in Sects.2.2.1,2.2. The data thus obtained should lead to improvements in the design of the loading devices, i.e., the exclusion or reduction of undesirable displacements.

The opposed-beam configuration, with the recording medium fixed to the object under investigation, is another efficient means of compension for the overall object displacement [2.44,45]. Here, because the recording medium follows all displacements of the strained object, the holographic interferogram will contain only information on the strain-induced displacements. It is recommended that the hologram be mounted on a nondeforming or weakly deforming part of the object. Another technique was proposed by *Shtan'ko* et al. [2.46]. They suggested that the object be illuminated via a flexible light guide, with one end connected to the laser and the other to the hologram holder.

In the case of small rigid-body displacements, where $|D_0| \simeq |d|$, their effect can be evaluated in a quantitative analysis of the fringe patterns. The various methods of holographic-interferogram interpretation are discussed below. They can involve a so-called *witness*, i.e., a nondeformable plate or body of any other appropriate shape which is mounted, as circumstances allow, on a nondeformable part of the object. One can also employ a nondeformable part of the object itself as a witness provided that it is of sufficient size. Quantitative analysis of the fringes observed on the witness yields information on the rigid-body displacement, while the fringe pattern on the object surface shows the sum of the strain-induced and overall displacements. The difference between these quantities thus gives the displacements due only to the strain.

Rigid-body displacements of an object can also be compensated for by using the so-called local-reference-beam configuration [2.47]. A version of this configuration is depicted in Fig.2.22. The object O under study is illuminated with light from a coherent point source S. Part of the illuminating beam falls on a mirror M fixed to the object and is reflected onto the hologram H forming a reference wave. In this case, the displacements are

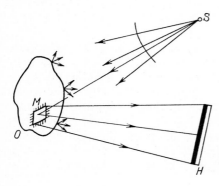

Fig.2.22. Interferometer setup with a local reference beam

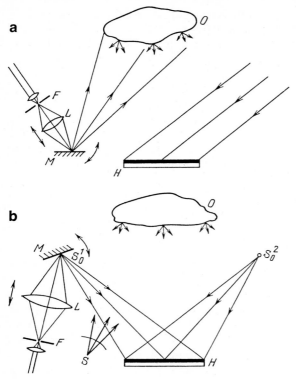

Fig.2.23a,b. Fringe compensation in real-time holography (a), fringe compensation in two-exposure holography (b)

compensated for by adjusting the reference-beam path. In another version of this method, part of the light illuminating the object is focussed onto a spot on the object surface [2.48]. The light scattered from this point produces a reference wave. Since the scattered wave is modulated by speckle, this technique is called *speckle reference beam holography*. The effect of the speckle pattern in the reference beam can be limited by reducing the size of the focussed spot on the object surface. The advantage of this arrangement is that no mirrors are needed. Large rigid-body displacements where $|\mathbf{D}_0| > |\mathbf{d}|$, can be compensated for during observation by using various methods of fringe control [2.49, 50].

Figure 2.23a displays an optical arrangement used to obtain real-time interferograms. The laser beam passing through a spatial filter F is focussed onto a point on the surface of a mirror M by a lens L. The spherical wave thus created illuminates the surface of the object O under study. The wave reflected from the surface impinges on the hologram H illuminated by the reference wave. After a first exposure the hologram is processed and returned into the optical system. When viewing the object through the hologram, the frequency and shape of the fringes can be changed by rotating the mirror M and displacing the lens L, i.e., by varying the phase of the wave reflected from the object surface. Since the fringes are observed in real time, one can readily determine the directions in which one should turn the mirror and displace the lens in order to reduce or completely eliminate the fringes on the nondeformable parts of the object or on the witnesses. Rigid-body displacements can be compensated for at each loading step.

An optical arrangement providing fringe shape and frequency control in two-exposure holographic interferometry is illustrated in Fig.2.23b. The object O is illuminated by the point source S. Two reference waves are produced by means of two point light sources S_0^1 and S_0^2, the reference wave from S_0^1 being formed by means of the spatial filter F, and lens L, and the mirror M just as in the real-time technique. The initial state of the object is recorded on the hologram H in the first exposure using the reference source S_0^1. In the second exposure, the strained object is exposed to the reference source S_0^2. By simultaneously illuminating the hologram thus obtained with both reference sources, the fringe pattern can be observed. By properly varying the phase of the reference wave from S_0^1 by displacing the lens and tilting the mirror, one can compensate for rigid-body displacements over a substantial part of the object surface.

The above methods of controlling the shape and frequency of interference fringes (Fig.2.23) are based on introducing an additional phase shift of a simple form (linear, spherical, cylindrical) into the object or reference wave. (This is equivalent to the introduction of a wedge, spherical lens, or cylindrical lens).

These methods substantially broaden the potential of application of holographic interferometry in solid mechanics. This is particularly useful in studies of deformation of low-rigidity structural elements. Figure 2.24a presents an interferogram of an area near a circular hold cut in a clamped plate acted upon by a bending moment of 0.044 N·m. This pattern does not permit evaluation of the effect of the hole on the deformation of the plate since the perturbations of the fringes are too small. Such an evaluation would require application of much higher loads at which, however, the fringes would no longer be resolvable. The methods of fringe control described above provide a possibility of overcoming this difficulty. Figure 2.24b shows an interferogram of the same plate loaded by a moment that is

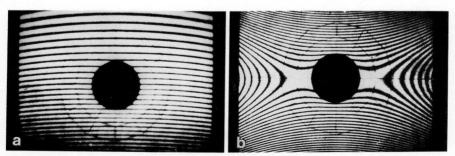

Fig.2.24a,b. Interferogram of plate with hole: (a) without compensation (bending moment - 0.044N·m), (b) with compensation (bending moment - 0.44N·m)

an order of magnitude higher, namely 0.44 N·m. This interferogram was obtained in the configuration of Fig.2.23b by introducing a linear phase shift which compensates for the rotation of the plate's middle section. The fringe pattern obtained can be used to determine the stresses along the edges of the hole.

Stimpfling et al. [2.51] succeeded in eliminating the effects of undesired displacements at the wavefront reconstruction stage by forming the reference wave with a flexible single-mode light guide which was rigidly fixed to the hologram holder. By varying the hologram position, the effects of in- and off-plane translations of the object by 5 and 3 mm, respectively, as well as rotations of 30 and 15 arc minutes about axes normal and tangential to the surface could be eliminated.

In sandwich holography, the compensation for translations and rotations of the object is achieved by rotating the sandwich. For the sandwich rotation angle ϕ required to compensate for the object rotation through an angle φ one can write [2.52]

$$\varphi = \tfrac{1}{2}\arctan\ell\{L[(n_c/\sin\phi)^2 - 1]\}^{-1/2} , \qquad (2.97)$$

where ℓ is the emulsion spacing in the sandwich, L is the sandwich-to-object distance, n_c is the refractive index of the glass substrate. Estimates show the angle ϕ to be two orders of magnitude larger than the angle φ, which is convenient for practical work.

For in-plane translation of an object, the angle ϕ required for compensation can be determined from [2.52]

$$|D_0| = L_S \ell[(L_S + L)(n_c/\sin\phi -1)^{1/2}]^{-1} , \qquad (2.98)$$

where L_S is the distance from the source illuminating the object to the latter. *Abramson* et al. [2.52] successfully demonstrated compensation of in-plane translation on the order of 1 mm. Compensation of rigid-body object translations in sandwich-holographic interferometry can also be achieved by varying the gap between the holograms and their relative position. Information on the in-plane object translation can be eliminated using spatial filtering in the Fourier plane of the light field reconstructed by the doubly exposed hologram [2.53, 54].

In temperature-induced deformations, rigid-body displacements may arise due to the motions of elements in thermal contact with the object under investigation. One should also bear in mind that when a layer of air (or any other medium) near the surface of interest is heated, its refractive index will change, thereby introducing an additional phase shift which can affect the fringe pattern. This effect, however, may be neglected if the temperature changes by a large amount, on the order of a few degrees [2.55, 56].

In conclusion, we note that ensuring the desired character of object deformation, which requires exclusion of rigid-body displacements affecting the fringe pattern, is a key problem in quantitative measurements by holographic interferometry. It is natural to assume that rigid-body displacements of an object can be reduced by reducing the applied loading. This consideration accounts for the widespread use of low-modulus materials. Experience shows, however, that holographic interferometry can also be employed at high levels of mechanical loading (up to 200 kN) [2.57] and also high temperatures (up to 800°C) [2.58].

2.3 Interpretation of Holographic Interferograms

2.3.1 Determination of the Displacement from Fringe-Localization Parameters and Contrast

One of the first methods for determining displacement by holographic interferometry involved the use of fringe-localization parameters and the spacing on the localization surface [2.59]. We choose a Cartesian coordinate system x_i $(i = 1, 2, 3)$ with the origin at the surface point of interest and the x_1 axis directed along the normal to the object surface. Let the interference fringes with the surface point viewed in the direction of the x_1 axis be localized at a distance x_{1B} from it. Then the tangential (in-plane) components X_2 and X_3 of the displacement vector can be found from the following expressions [2.59]

$$X_2 = \lambda x_{1B}/L_{x2} , \quad X_3 = \lambda x_{1B}/L_{x3} , \tag{2.99}$$

where L_{x2} and L_{x3} are the fringe spacings along the corresponding coordinate axes. To determine the component X_1, we have to find the displacements X_2' and X_3' at another direction of observation of the surface point in question. Note that the set (2.99) is similar to (2.80) which describes the localization conditions for the fringes viewed through a slit aperture.

Despite the apparent simplicity of (2.99), this method of fringe interpretation did not enjoy widespread recognition in practice because of the insufficiently high accuracy in the fringe localization position, particularly in cases where the fringes are localized close to the surface image.

Information on the surface-point displacements of a strained object can also be derived from an analysis of the fringe contrast on the surface. For a circular aperture of diameter D with angular dimension less than the scattering indicatrix of roughness on the object surface, one can write the following expression for the correlation length l_0 [2.60]:

$$\ell_0 \simeq 0.6\lambda L_0/D$$

where L_0 is the distance from the point of observation to the object.

By viewing the fringe pattern on the object surface through a variable-aperture optical system, one can see the pattern disappear when the in-plane component of displacement, e.g., X_2 becomes larger than the correlation length:

$$X_2 > \ell_0 .$$

This technique of interferogram processing can be improved when two components of the displacement vector, X_2 and X_3 can be determined. Obviously, this can be achieved by viewing the fringe pattern through a variable aperture which rotates about the x_1 axis. By placing the aperture in the position of maximum fringe contrast, one determines the direction of the displacement component normal to the viewing axis. Next, by varying the aperture until the fringes disappear, one measures the magnitude of this component. This method can be used only under the conditions where the fringe patterns are of high quality and the noise in the reconstructed image is sufficiently low to permit accurate measurement of the fringe characteristics, and reliable observation of the appearance and disappearance of interference fringes in the area of interest.

The methods based on measuring the fringe contrast have recently witnessed new developments. One of them will be discussed in Sect. 4.2.3.

2.3.2 Interferogram Interpretation Based on the Absolute Fringe Order

The principal relation of holographic interferometry (2.47) permits determination, at an arbitrary point on the object surface, of the projection of the displacement vector on the sensitivity vector. To find the total displacement vector at a point, one must determine three projections onto different sensitivity vectors. In other words, it is required to record interference fringes from three different viewing directions. By properly processing three interferograms for a given surface point, one arrives at the system of couples equations

$$\mathbf{k}_1 \cdot \mathbf{d} = n_1 \lambda \,,$$
$$\mathbf{k}_2 \cdot \mathbf{d} = n_2 \lambda \,,$$
$$\mathbf{k}_3 \cdot \mathbf{d} = n_3 \lambda \,, \tag{2.100}$$

where n_1, n_2, n_3 are the *absolute fringe orders* at the point of interest determined from the corresponding interferogram, $\mathbf{k}_1$, $\mathbf{k}_2$, $\mathbf{k}_3$ are the sensitivity vectors for the three chosen viewing directions, and λ is the light wavelength. Solving these coupled equations for noncoplanar sensitivity vectors yields the displacement vector $\mathbf{d}$. The key aspect which determines the applicability of this method is that the absolute fringe orders on all interferograms must be known. Since the fringe order is found by counting from the zero-order fringe, this method is sometimes called the *zero-motion fringe method*. This approach to interferogram interpretation was first proposed by *Ennos* [2.30] and later generalized by *Sollid* [2.31]. Its advantage is that the absolute fringe orders can be read out directly from photographs of fringe patterns. To determine the displacement field, this procedure is repeated for all points of interest on the object surface.

The coupled equations (2.100) can conveniently be represented in terms of the vector components in the Cartesian coordinate system x_1, x_2, x_3 on the object under study. The system of linear equations to determine

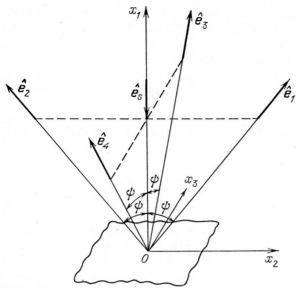

Fig.2.25. Interferometer configuration with symmetrical viewing directions

the displacement vector components X_1, X_2, X_3 at the point of interest has the form

$$KX = \lambda N \qquad (2.101)$$

where $X = [X_1, X_2, X_3]^T$ is the displacement vector, $N = [n_1, n_2, n_3]^T$ is the absolute fringe order vector, and K is a 3×3 *holographic-interferometric sensitivity matrix*, its elements being equal to the projections of the sensitivity vectors on the coordinate axes.

The determination of the displacement-vector components for objects with plane surfaces can be greatly facilitated by using the interferometer setup shown in Fig.2.25 [2.61-64]. The object is illuminated in a direction specified by the vector $\hat{e}_s$ normal to the surface, and the four viewing directions are symmetrically spaced about it. The x_1 axis of the x_1, x_2, x_3 coordinate system lies along the normal to the surface at point O. The unit observation vectors $\hat{e}_1$ and $\hat{e}_2$ lie in the $x_1 x_2$ plane, the vectors $\hat{e}_3$ and $\hat{e}_4$, in the $x_1 x_3$ plane. The angle ψ between the viewing direction and the normal to the surface determines the interferometer sensitivity with respect to the in-plane displacement-vector component.

For the two viewing directions defined by the vectors $\hat{e}_1$ and $\hat{e}_2$ the system (2.101) can be written as

$$\begin{bmatrix} 1+\cos\psi & -\sin\psi \\ 1+\cos\psi & \sin\psi \end{bmatrix} \begin{bmatrix} X_1 \\ X_2 \end{bmatrix} = \lambda \begin{bmatrix} n_1 \\ n_2 \end{bmatrix}, \qquad (2.102)$$

where n_1, n_2 are the absolute fringe orders at the point O for the viewing directions $\hat{e}_1$ and $\hat{e}_2$, respectively. Solving (2.102) yields

$$X_1 = \lambda(n_1 + n_2)[2(1 + \cos\psi)]^{-1} \, , \qquad (2.103)$$

$$X_2 = \lambda(n_2 - n_1)(2\sin\psi)^{-1} \, . \qquad (2.104)$$

Similar reasoning for the viewing directions specified by the vectors $\hat{e}_3$ and $\hat{e}_4$ yields the following expressions for the displacement components X_1 and X_3:

$$X_1 = \lambda(n_3 + n_4)[2(1 + \cos\psi)]^{-1} \, , \qquad (2.105)$$

$$X_3 = \lambda(n_4 - n_3)(2\sin\psi)^{-1} \, , \qquad (2.106)$$

where n_3 and n_4 are the absolute fringe orders at the point O for the viewing directions $\hat{e}_3$ and $\hat{e}_4$, respectively. The criterion of validity for all the displacements thus found is the coincidence of the values of X_1 obtained by means of (2.103, 105). Equations (2.103–106) can be also used for other surface points but only under collimated illumination and if a telecentric system is used for observation.

The interpretation of interferograms involving determination of absolute fringe orders can be carried out in both a single- and multi-hologram interferometer setup.

In the Leith-Upatnieks arrangement the object is illuminated by a spherical or a plane wave, and the fringe patterns required for the determination of the displacement components are viewed from the three directions within the hologram aperture [2.65]. The optical arrangement of such an interferometer is shown in Fig.2.26. The direction of illumination of the point Q is specified by the unit vector $\hat{e}_S$, and that of observation by the unit vectors $\hat{e}_1$, $\hat{e}_2$, and $\hat{e}_3$. The sensitivity with respect to the in-plane displacement-vector components can be enhanced by increasing the angles between the viewing directions. This can be achieved by either increasing the size of the hologram proper or by bringing it closer to the object surface. However, this may entail technical difficulties in illumination of the object and in formation of the reference wave. The angles between the viewing

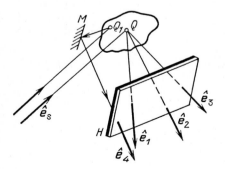

Fig.2.26. Single-hologram interferometer for fringe interpretation based on absolute order numbers

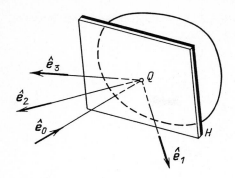

Fig.2.27. Single-hologram interferometer using reflection hologram

directions in a single-hologram arrangement can be increased by using a mirror near the object [2.66]. Figure 2.26 shows the rays from the light source, the point Q_1 on the object surface, and a mirror M in the viewing direction specified by the vector $\hat{e}_4$.

The opposed-beam optical arrangements offer broad possibilities for the measurement of displacements [2.67-69]. A configuration which can be used to reconstruct the image recorded on such a hologram by illuminating it with a plane wave in the direction of the vector $\hat{e}_0$ is shown in Fig.2.27. Due to the small distance between the hologram and the object surface, for the same hologram dimensions as in the Leith-Upatniek's arrangement, one can obtain larger angles between the viewing directions $\hat{e}_1$, $\hat{e}_2$, and $\hat{e}_3$.

When one changes the viewing direction in the above single-hologram interferometric setups, both shape and dimensions of the object image on different fringe patterns will change. This creates difficulties in the identification of object points on different interferograms corresponding to different directions of observation. To solve this problem, one usually draws a grid on the object surface, and determines the displacement of the grid intersections. The difficulty can be avoided, however, by using a single-hologram interferometer with one direction of observation $\hat{e}_1$ of the surface point under study and three directions of illumination specified by the vectors $\hat{e}_1^S$, $\hat{e}_2^S$ and $\hat{e}_3^S$ [2.70, 71] (Fig.2.28). To each object wave corresponds a unique reference wave. In order for each object wave to interfere only with its reference, the path difference of the object beams is made larger than

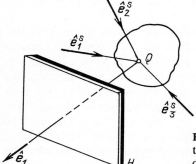

Fig.2.28. Single-hologram interferometer with three directions of illumination and one viewing direction

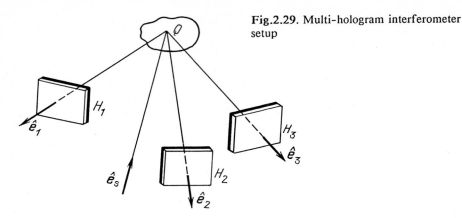

Fig.2.29. Multi-hologram interferometer setup

the coherence length of the laser light. Thus, one simultaneously records on one photographic plate three doubly-exposed holograms. The fringe pattern is observed by sequentially reconstructing the light waves recorded on the hologram with the corresponding reference waves and the viewing direction is held fixed, which ensures the absence of any parallax in the reconstructed images. When writing the coupled equations (2.101) for the determination of the displacement vector, one must, therefore, take into account that only the direction of illumination is varied while the viewing direction is fixed. This technique entails a substantial loss of light signal and therefore requires the use of high-power lasers. This, as well as the complexity of the optical systems utilized in such interferometers, represents a serious obstacle to widespread application of the method in deformation studies.

In the multi-hologram technique [2.72], for each of the three viewing directions $\hat{e}_1$, $\hat{e}_2$, $\hat{e}_3$ of the surface point Q, an individual hologram, H_1, H_2 and H_3 is provided (Fig.2.29). Despite technical difficulties encountered in the construction of the optical system, one can obtain practically any desired geometry of the sensitivity vectors. Object points on different interferograms are likewise identified, as in the single-hologram method, by plotting a grid. It is possible to project the real image on the object surface and to record interferograms from one viewing direction [2.73]. In this case the real image of the object is reconstructed by illuminating the holograms placed at the point of exposure with an unexpanded laser beam. This approach does not entail any distortions in the shape or dimensions of the image.

The geometrical parameters of the interferometer arrangement are determined, as a rule, by measuring the positions of the illumination source and of the observation points relative to a surface point which is taken as the origin. If we know the geometry of the object, we can compute the components of the illumination and observation vectors for other points of the object surface which, in their turn, permit construction of the sensitivity matrices. For small object surfaces the processing of the interferograms can be facilitated by using collimated light for the illumination. Under these conditions the unit illumination vector $\hat{e}_s$ will be constant over the entire object surface.

70

Another solution of this problem suggested by *Pryputniewicz* [2.74] is to obtain a series of doubly-exposed holograms of the surface of the object which undergoes known displacements (translations or rotations) between the exposures. By knowing the displacement components and the fringe orders at the points of interest, we can determine the sensitivity matrices at these points by using (2.101). Now, if the object under study has suffered a deformation while the parameters of the optical setup are kept fixed, then the unknown displacements can be determined from the measured fringe orders and sensitivity matrices. This method is quite efficient when mirrors are used to increase the angles between the viewing directions (Fig.2.27).

2.3.3 Determination of Fringe Order and Sign

The numbers and absolute orders of interference fringes are determined by counting them starting from the *zero-motion fringe*. As follows from the principal relation (2.41), a zero phase difference corresponding to the zero-order fringe is reached either when the displacement vector is zero ($d \equiv 0$) or when it is orthogonal to the sensitivity vector ($d \perp k$). We first consider the major techniques used to find the position of the zero-motion fringe in the two-exposure method.

The simplest case is when there are areas on the object surface which are known not to have undergone displacement under loading. As shown in Sect.2.2.1, bright interference fringes will pass through these areas. In the interferogram shown in Fig.2.1a, for example, the zero-motion fringe lies near the fixed end of the beam. In some cases it may turn out to be appropriate to deliberately build in unstrained regions in the object, under the condition that they will not affect the desired deformation. Among this class of problems is the determination of displacements and deformations of small areas on a part with holes, notches, or similar features.

Another method of zero-motion fringe identification makes use of the fact that the zero-motion fringe does not change position as the viewing direction within the hologram aperture or the direction of object illumination is varied [2.30]. In another version of this method, two interferograms of the object surface are obtained for different directions of illumination and one fixed viewing direction. When the two interferograms are superimposed, the zero-motion fringes should coincide. Diffuse illumination can be used to visualize the zero-motion fringe [2.75, 76].

The position of unstrained areas on the object surface can also be determined by time-average interferometry. Irrespective of the actual character of the object deformation, the interference fringes corresponding to a zero-displacement region have the maximum brightness. *Köpf* [2.77] suggested a method for determining the zero-motion fringe position based on a combined application of the two-exposure and time-average techniques. Because this method is of practical importance, we consider it in more detail.

Suppose that in a first exposure of duration τ_1 we record the original state of the object surface and denote its phase by ϕ. Next the object is loaded, and a hologram is obtained during the time τ_2 in such a way that the rate of phase change remains constant. The object wave phase at a time t will then be

$$\phi + (\delta/\tau_2)t \, ,$$

where δ is the maximum phase difference between the exposures. Finally, the deformed object in steady state is subjected to a third exposure of duration τ_1 and at an object-wave phase $\phi+\delta$. In order to determine the intensity distribution of the reconstructed virtual image by means of the triply-exposed hologram thus obtained, we can represent the integral in (2.33) as a sum of three integrals to find the characteristic fringe function

$$|\mathbf{M}(\delta)|^2 = \left[2\tau_1 \cos\delta/2 + \tau_2 \frac{\sin\delta/2}{\delta/2} \right]^2 . \qquad (2.107)$$

This expression describes the interference fringes produced in a combined application of the two-exposure and time-average methods. For $\tau_2 = 0$ we will have the fringe function of the two-exposure method, and for $\tau_1 = 0$, that of the time-average technique with the phase being varied at a constant rate (Table 2.1). As the ratio $\tau_2/2\tau_1$ grows, the intensity of the zero-motion fringes will increase relative to the other bright fringes. At the same time the roots of (2.107) and of the function $\cos^2 \delta/2$ will become slightly different, however, these differences may by neglected for $\tau_2/2\tau_1 \simeq 2$.

Thus, the above three-exposure method allows one to obtain holographic interferogram similar to a doubly-exposed hologram but with the zero-motion fringes more intense than the other bright fringes. Figure 2.30 shows a triply-exposed interferogram ($\tau_2/2\tau_1 = 2$) of a cylindrical shell under torsional stress, which clearly reveals a bright zero-motion fringe. One may compare it with the interferogram in Fig.2.21d of the same shell placed in the same loading conditions but obtained by the two-exposure technique. An important advantage of the three-exposure method is that it allows observation not only of the areas free of deformation (for instance, the region on the left in Fig.2.30 where the shell is fixed), but also of the parts of the object surface where the sensitivity vector is orthogonal to the displacement vector. The latter is clearly evidenced by the bright fringe running along the cylinder generatrix.

In cases where the reflection holograms are fixed to a surface point of the strained object, the zero-motion fringe will always be in the vicinity of this point.

When zero-motion fringes are not seen in an interferogram, the fringe orders can be determined by means of elastic strings connecting an object which will not deform with the surface of the object under study [2.78]. The fringes are then counted from the zero-motion fringe located at the point where the string is fixed to the nondeforming body. For correct fringe-order readout, the fringes at the point where the string joins the surface of the object under study must be continuous. The interferograms presented in Fig.3.12 show such a string made of rubber.

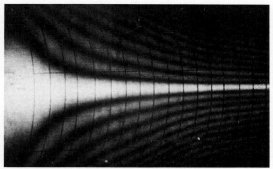

Fig.2.30. Three-exposure interferogram of a cylindrical shell subjected to torsion

The solution of the coupled equations (2.100) defining the displacement vector for a surface point will also depend on the choice of the fringe-order sign, $\pm n_1$, $\pm n_2$, $\pm n_3$, since the fringe orders can be read out on the positive or negative side from the zero-motion fringe. The assignment of the sign follows a convention which is sometimes not straightforward. In some instances photographs of fringe patterns may turn out to be sufficient for a correct sign assignment whereas under some conditions, an analysis of the fringe pattern variation with varying viewing direction within the hologram aperture is required. It is essential that the fringe-order signs be assigned by the same rule for all viewing directions.

One approach that was proposed for eliminating this ambiguity in the two-exposure method involves an interferometer configuration with one direction of illumination and three viewing directions [2.79,80]. The interference pattern generated by a given surface point in the initial and displaced states is known to represent a family of hyperboloids of revolution with the axis passing through the two positions of the point in question. At sufficiently large distances, the hyperboloids may be approximated to a high degree of accuracy by cones. If the direction from which the point is observed lies on the surface of one of these cones, then the fringe passing through it will always be of the same order. For the zero-order fringe a so-called "zero cone" should exist. The viewing directions lying on the opposite sides of the zero cone yields, for the point in question, fringes of opposite signs. Thus, to assign signs to fringe orders, one must establish the position of the zero cone. In a multi-hologram interferometer setup one can find the directions in which the fringe order decreases at fixed points on the object surface by varying the viewing directions within the aperture of each hologram. Such directions specify the position of the zero cone. For instance, in the particular case where on all holograms such changes of the viewing direction occur in the same direction, all fringes at the point in question have the same sign. By varying the viewing direction within the hologram aperture one can also establish the sites where the interference fringes are generated. At these sites the fringe orders acquire extremum values.

The fringe-order signs can be also determined by calculation. One of the methods employs an overdetermined system of equations (2.100) with

73

the number of coupled equations j greater than 3 [2.81]. The possible number of fringe-order sign variations in this case is 2^{j-1}. One variation of the fringe-order signs is chosen and used to construct all possible combinations of the three equations. The largest difference between the vectors of the solution thus obtained is denoted by $\Delta\mathbf{d}_{max}$. This procedure is repeated for the rest of the variations of fringe signs in the original overdetermined system of equations. The smallest of the values will correspond to the correct choice. Obviously, only a computer is capable of coping with such a large volume of calculations. *Harnisch* et al. [2.82] proposed another approach in which a brute force search is made through the various fringe-order signs for an overdetermined system of equations, the displacement vector of interest being the one corresponding to the minimum standard deviation.

At least one additional viewing direction should be added to the required three in order to determine the correct fringe-order combination [2.83]. Having three equations in (2.100), one can, by varying the fringe-order signs, obtain four sets of the displacement vector components. By sequentially substituting each of the sets into the fourth equation corresponding to the additional direction of observation, one can find the displacement components which make it an identity.

In the real-time method the ambiguity in the fringe-order sign is eliminated by observing the fringe shift under continuous loading of the object [2.84]. In cases where the interferogram contains the zero-order fringe, a change in the loading makes the higher-order fringes move toward or away from it. The interference fringes are then given the same sign. There may be situations where the fringes shift in one direction relative to the zero-motion fringe, in which case the fringe-order signs on either side of the zero fringe are opposite. In the absence of the zero-motion fringe, fringe "sinks" will be observed. The shift of the fringes toward these sinks under loading can also permit identification of the fringe-order sign.

Because the recorded phase difference is an argument of the squared characteristic fringe function, the solution of (2.100) can yield only the magnitude of the displacement vector to within the sign. To determine the sign, i.e. its direction one must use a priori information on the character of deformation of the object at its characteristic points or employ other experimental methods of finding the displacement components at reference points of the object surface.

The fringe numbers at grid intersections on the object surface can be found visually. The accuracy of their determination by linear interpolation is on the order of 0.2 fringe. The use of microphotometers improves the accuracy by a factor of two. However, an accuracy of up to 0.001 fringe can be reached when special optical setups with electronic phase-measuring devices are used [2.85-87].

To illustrate the above discussion, Fig.2.31 presents a holographic image of a cylindrical shell with a rectangularly cutout subjected to a torsional load and obtained by the three-exposure method. One can see the position of zero fringes around the cutout and the extremal values of the fringe or-

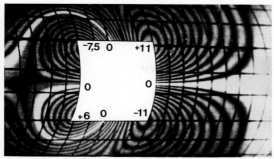

Fig.2.31. Three-exposure interferogram of a cylindrical shell with a rectangular notch subjected to torsion

ders. The fringe order reverses its sign as one crosses the zero motion fringe.

2.3.4 Interpretation of Interferograms
Based on the Relative Fringe Order

Aleksandrov and *Bonch-Bruevich* [2.29] proposed a method for interpreting holographic interferograms, which does not require determination of the zero-order fringe. It involves the measurement of the shift with respect to the object-surface image that the fringes undergo as the direction of observation of the point under consideration is varied within the hologram aperture.

For a point on the object surface we write the principal relation of holographic interferometry (2.44) consecutively for two viewing directions specified by unit vectors $\hat{e}_1$ and $\hat{e}_2$

$$(\hat{e}_1 - \hat{e}_S)\cdot\mathbf{d} = \lambda n_1 , \qquad (2.108)$$

$$(\hat{e}_2 - \hat{e}_S)\cdot\mathbf{d} = \lambda n_2 , \qquad (2.109)$$

where n_1 and n_2 are the absolute fringe orders at the point in question for the viewing directions specified by the unit vectors $\hat{e}_1$ and $\hat{e}_2$, respectively; $\hat{e}_S$ is the unit vector of illumination. Subtracting (2.108) from (2.109) yields

$$(\hat{e}_2 - \hat{e}_1)\cdot\mathbf{d} = \lambda(n_2 - n_1) . \qquad (2.110)$$

Equation (2.110) thus does not depend on the illumination conditions. We denote the fringe-order difference $n_2 - n_1$ at this point by n_{21}, and call it the *relative fringe order*. In practice, the relative order of a fringe is determined by counting the number of the fringes that pass through the point as the viewing direction is varied continuously from $\hat{e}_1$ to $\hat{e}_2$. The relative fringe order depends on the magnitude and direction of the displacement vector and on the actual geometry of the fringe observation. Equation (2.110) can thus be rewritten in the form

75

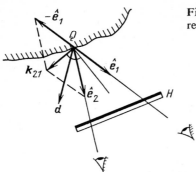

Fig.2.32. Locating the sensitivity vector k_{21} in the relative fringe order counting method

$$(\hat{e}_2 - \hat{e}_1) \cdot d = \lambda n_{21} . \qquad (2.111)$$

We will call the difference of the two unit viewing vectors in (2.111) the sensitivity vector in the relative fringe-order method. We denote it by k_{21}, the two indices corresponding to the two directions of observation:

$$k_{21} = \hat{e}_2 - \hat{e}_1 . \qquad (2.112)$$

Equation (2.111) can now be written in terms of the sensitivity vector in the following way:

$$k_{21} \cdot d = \lambda n_{21} . \qquad (2.113)$$

Without loss of generality, we will assume that the displacement vector d lies in the plane defined by the unit vectors $\hat{e}_1$ and $\hat{e}_2$. Now one can readily see from the configuration in Fig.2.32 that the sensitivity vector is orthogonal to the bisector of the angle between the viewing directions. In the particular case where the viewing directions $\hat{e}_1$ and $\hat{e}_2$ are symmetric with respect to the normal to the object surface, the sensitivity vector is tangent to the surface at the point of interest and, hence, the sensitivity with respect to the out-of-plane component of the displacement vector is zero.

To determine the displacement vector for a point on the object surface, we must write coupled equations of the type in (2.113) for three variations of the viewing direction:

$$k_{21} \cdot d = \lambda n_{21} ,$$
$$k_{43} \cdot d = \lambda n_{43} ,$$
$$k_{65} \cdot d = \lambda n_{65} , \qquad (2.114)$$

where n_{21}, n_{43}, n_{65} are the relative orders, and k_{21}, k_{43}, k_{65} are the sensitivity vectors for the corresponding variations of the viewing directions. The solution of the set (2.114) for the noncoplanar sensitivity vectors yields the displacement vector for the object point at the surface. It should be pointed out that in the relative fringe-order method all information re-

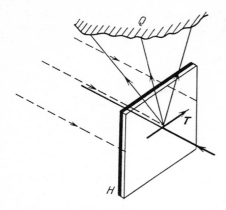

Fig.2.33. Determination of relative fringe order at a point by scanning doubly exposed hologram with a narrow laser beam

quired to determine the displacements is derived directly from the reconstructed image.

Relative orders for a point on the object surface can be also determined by observing the virtual image. This procedure is very tedious, however, particularly in the cases where the number of the fringes passing through the point in question is small and fractional orders have to be measured.

Relative fringe orders may be conveniently found by using the real image observed by scanning the hologram with a narrow laser beam [2.88-90]. Figure 2.33 displays a doubly exposed hologram obtained with a plane reference wave that is reconstructed by illumination with a narrow (unexpanded) laser beam which is conjugate to the reference beam used to record the hologram H. The original reference beam is shown in the figure by the dashed line. The illuminated part of the hologram produces a real image of the object with superimposed interference fringes. Because the diameter of the illuminating beam is small, the interference fringes characterizing the deformation of the object are observed together with the real image of the object surface. A shift of the laser beam in the direction of the vector T will change the viewing direction, and as a result, interference fringes will pass through the surface point Q. The number of fringes passing through the point gives the relative order. A notable feature of this technique is that the viewing direction is fixed with high accuracy. Because a real image is used, it is also essential that the process of relative fringe-order determination be easily automated. This can be achieved by placing a photodiode at the surface point in question to record the variation of the light intensity as the fixed hologram is scanned by an unexpanded laser beam. By analysing the dependence of the intensity on the variation of the viewing direction, one can determine the relative fringe order. This method has the disadvantage that a decrease of the illuminated area on the hologram increases the speckle size in the real image which can make fringe counting difficult, particularly where the displacement gradients are large.

The assignment of the sign of the relative fringe order is purely conventional and can be made, for instance, by the following rule: if the fringes shift in the direction in which the viewing position is varied, then

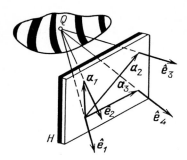

Fig.2.34. One-hologram interferometer for fringe interpretation based on relative fringe order readout

the fringe orders are assigned a positive sign, and if they move in the opposite direction then they are considered to be negative.

Just as in the method involving absolute fringe orders, solving the coupled equations (2.114) yields the displacement vector of the surface point in question to within its sign. The sign of the displacement vector is determined either by considering the physical meaning of the problem and taking the points where the direction of the displacement is known a priori, or by using other experimental techniques to measure the displacement. To determine the displacement field, the above procedure must be repeatedly applied for different points on the object surface.

As in the absolute-order method, interpretation of interferograms on relative fringe orders is possible in both single- and multi-hologram interferometer configurations. The most widely used single-hologram interfometer setup is arrangment due to Leith and Upatnieks [2.91-93] for obtaining a doubly-exposed hologram. Observation of the virtual image of an object is illustrated in Fig.2.34. All the changes in the viewing direction necessary to construct the system (2.114) for the point Q on the object surface are confined within the aperture of the hologram H. The four final directions of observation denoted by the unit vectors $\hat{e}_1$, $\hat{e}_2$, $\hat{e}_3$ and $\hat{e}_4$ define the various means of determination of the relative fringe orders. We denote their projections in the plane of the hologram for the chosen changes in the viewing direction by the vectors a_1, a_2 and a_3. In this interferometer setup all the changes are made from the same original viewing direction $\hat{e}_1$.

In another modification of the relative fringe-order method used in a single-hologram interferometer configuration, the viewing direction is changed in such a way that the relative fringe order be zero for a fixed point on the object surface [2.94-97]. Such a point is called a pole. It can be shown that in this case, the viewing directions lie on the surface of a circular cone, with its axis coinciding with the direction of the displacement vector. This cone intersects the hologram along the so-called κ-line.

The essence of this method is illustrated in Fig.2.35. We choose the pole to lie at the point O on the object surface and let a whole fringe pass through it. We fix a κ-line on the hologram H for this pole, and on it, three points of observation B_1, B_2 and B_3. The fourth observation point should lie off the κ-line. Four fringe patterns are recorded from these viewing positions. In Fig.2.35 all the three fringes passing through the pole when

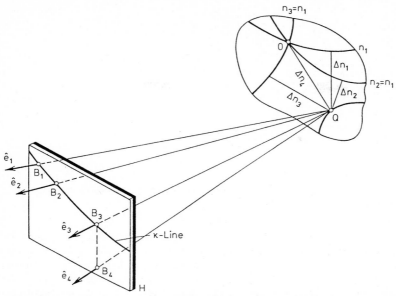

Fig.2.35. Interferogram interpretation based on the κ-line in the relative fringe order counting technique

observed from the directions specified by the points B_i $(i = 1, 2, 3)$ are shown at the same time. We now find the relative order $n_{\kappa 4}$ of the fringe at the pole as one passes from the κ-line to the point B_4. Photographs made for an arbitrary point Q on the object surface can be used to count the number of interference fringes between this point and the pole, which will be denoted by Δn_{i0} $(i = 1, 2, 3, 4)$. In this case, the displacement vector **d** will be determined from the following system of equations

$$(\hat{\mathbf{e}}_1 - \hat{\mathbf{e}}_2) \cdot \mathbf{d} = \lambda(\Delta n_{10} - \Delta n_{20}) \,,$$

$$(\hat{\mathbf{e}}_1 - \hat{\mathbf{e}}_3) \cdot \mathbf{d} = \lambda(\Delta n_{10} - \Delta n_{30}) \,,$$

$$(\hat{\mathbf{e}}_1 - \hat{\mathbf{e}}_4) \mathbf{d} = \lambda[\Delta n_{10} - (\Delta n_{40} + n_{\kappa 4})] \,,$$

where $\hat{\mathbf{e}}_j$ $(j = 1, 2, 3, 4)$ are the unit vectors of the directions of observation of point Q from the points B_i $(i = 1, 2, 3, 4)$, respectively. To find the displacement vector for an arbitrary surface point one must determine the relative order $n_{\kappa 4}$ of the fringe at the pole when crossing over from the κ-line to the observation point B_4 only once.

In practice, it may be advantageous to expand the range in which the angle of observation varies with the aim of obtaining large relative fringe orders. When using the arrangement in Fig.2.35, this problem can be solved by either increasing the dimensions of the hologram or bringing it closer to the object. This, however, may cause difficulties in the formation of the object and reference waves. The range of the viewing-angle variation can

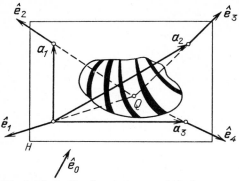

Fig.2.36. Reflection hologram interferometer configuration for fringe interpretation based on relative fringe order readout

be substantially increased by using a reflection-hologram interferometer. Figure 2.36 illustrates the reconstruction of the light waves recorded on a hologram H by illumination with a plane wave in the direction of the unit vector $\hat{e}_0$. The hologram was obtained by the two-exposure technique with the photographic plate fixed close to the object surface. The final directions of observation are specified by the unit vectors $\hat{e}_1, \hat{e}_2, \hat{e}_3$ and $\hat{e}_4$, and the vectors $\mathbf{a}_1, \mathbf{a}_2, \mathbf{a}_3$ on the hologram refer to the directions in which the viewing position is varied.

It may be useful to employ reflection holograms in an arrangement where the viewing position is varied in a circle, thus permitting separate determination of the in-plane and out-of-plane components of the displacement vector. [2.98, 99]. Consider the arrangement in Fig.2.37 for observing an image which is reconstructed when the hologram H is illuminated with a reference wave in the direction of the unit vector $\hat{e}_0$ [2.98]. The x_1-axis of a Cartesian coordinate system is along the normal to the ob-

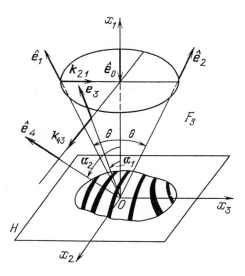

Fig.2.37. Reflection hologram interferometer setup with observation point moved along a circle

ject surface at the point of interest, and the x_2 and x_3 axes are tangent to the surface. Let us consider the fringe moves in the direction in which the point O is observed is varied in a circle with the center on the normal to the object surface. If the fringes stop at the point O when viewed from the direction specified by $\hat{e}_1$, we have reached a position where the displacement vector lies in the plane defined by the normal to the object surface and $\hat{e}_1$. The next position at which the fringes will stop is after the viewing direction has been rotated by 180°. This direction is defined in Fig.2.37 by the vector $\hat{e}_2$. For the viewing vectors $\hat{e}_1$ and $\hat{e}_2$, the sensitivity vector k_{21} is orthogonal to the normal to the object surface and, hence, the sensitivity with respect to the out-of-plane component of the displacement vector will be zero. This conclusion is valid for any pair of viewing vectors emerging from points lying on the same diameter of the circle. It can be readily shown that if the number of fringes which passed through the point O when the viewing direction varied from $\hat{e}_1$ to $\hat{e}_2$ is n_{21}, then the in-plane component of the displacement vector X_3 can be found by means of the expression

$$X_3 = \lambda n_{21}(2\sin\theta)^{-1} , \qquad (2.115)$$

where θ is the angle between the viewing direction and the x_1-axis. The relative fringe order at the point in question depends on the angle θ, with n_{21} being larger at larger angles.

We now let the viewing direction lie in the plane orthogonal to that defined by the unit vectors $\hat{e}_1$ and $\hat{e}_2$. In Fig.2.37 this direction is specified by the vector $\hat{e}_3$. By varying it from $\hat{e}_3$ to $\hat{e}_4$, we can determine the relative fringe order n_{43} at the point O. The sensitivity vector k_{43} will then be orthogonal to the in-plane component of the displacement vector X_3. Hence, to determine the out-of-plane component of the displacement vector X_1 we have

$$X_1 = \lambda n_{43}(\cos\alpha_1 - \cos\alpha_2)^{-1} , \qquad (2.116)$$

where α_1 and α_2 are the angles that the viewing directions $\hat{e}_3$ and $\hat{e}_4$ make with the x_1-axis.

It is instructive to apply the above method to determining the absolute fringe order at a given point on the object surface [2.98]. Writing the principal relation of holographic interferometry (2.44) for the viewing directions $\hat{e}_3$ and $\hat{e}_4$ and substituting into them the out-of-plane component of the displacement vector X_1 from (2.116), one obtains

$$n_1 = n_{43} \frac{\cos\alpha_1 + 1}{\cos\alpha_1 - \cos\alpha_2} , \qquad (2.117)$$

$$n_2 = n_{43} \frac{\cos\alpha_2 + 1}{\cos\alpha_1 - \cos\alpha_2} . \qquad (2.118)$$

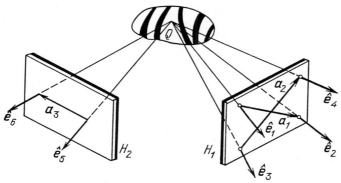

Fig.2.38. Two-hologram interferometer setup for interferogram interpretation based on relative fringe order counting

The absolute fringe order thus found can also be used to determine the fringe orders for other points on the object surface, thus opening a way to interferogram interpretation based on absolute fringe orders.

Multi-hologram interferometer setups broaden the range of sensitivity vector variation over a given surface region and offer additional possibilities for more precise determination of the displacement vector components [2.100]. Figure 2.38 shows such an interferometer setup. The six viewing directions specified by the unit vectors $\hat{e}_i$ ($i = 1,2,3,...,6$) define three changes in the direction of observation necessary to construct the system of (2.114). The two changes defined by the vectors $\mathbf{a}_1$ and $\mathbf{a}_2$ are made in hologram H_1 and the change specified by $\mathbf{a}_3$, in hologram H_2.

2.3.5 Interferogram Interpretation Based on the Fringe-Order Difference

The method of interferogram interpretation proposed by *Vlasov* and *Shtan'ko* [2.101] requires determination of neither absolute nor relative fringe orders. Their idea is to use fringe-order differences for several points on the interferogram of the surface.

Consider the interferometer arrangement depicted in Fig.2.39 with a point illumination source S and one observation point 1B. The unit illumination vectors of two points on the object surface, Q_1 and Q_2, are denoted by $\hat{e}_1^S$ and $\hat{e}_2^S$, respectively and the unit viewing vectors for these points by $\hat{e}_1^{1B}$ and $\hat{e}_2^{1B}$. Here the superscript refers to the index of the observation point and the subscript to that of the surface point. The principal relation of holographic interferometry (2.44) can be used to write for each of the points Q_1 and Q_2

$$(\hat{e}_1^{1B} - \hat{e}_1^S) \cdot \mathbf{d}_1 = n_1^{1B} \lambda , \qquad (2.119)$$

$$(\hat{e}_2^{1B} - \hat{e}_2^S) \cdot \mathbf{d}_2 = n_2^{1B} \lambda , \qquad (2.120)$$

82

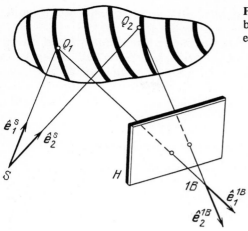

Fig.2.39. Interferogram interpretation based on fringe order number differences

where n_1^{1B} and n_2^{1B} are the absolute fringe orders for the points Q_1 and Q_2 determined by observing them from point 1B; $\mathbf{d}_1$ and $\mathbf{d}_2$ are the displacement vectors of the points Q_1 and Q_2. Subtracting (2.119) from (2.120) we obtain

$$(\hat{e}_2^{1B} - \hat{e}_2^{S}) \cdot \mathbf{d}_2 - (\hat{e}_1^{1B} - \hat{e}_1^{S}) \cdot \mathbf{d}_1 = (n_2^{1B} - n_1^{1B})\lambda . \qquad (2.121)$$

The absolute fringe-order difference between the points Q_1 and Q_2 for the observation point 1B, which is equal to the number of the fringes between these points, is denoted by Δn_{21}^{1B}. As follows from (2.121), one must obtain interferograms from six observation points in order to determine the vectors $\mathbf{d}_1$ and $\mathbf{d}_2$ at the points Q_1 and Q_2. To find the vectors $\mathbf{d}_1$ and $\mathbf{d}_2$ we write the following system of equations:

$$(\hat{e}_2^{1B} - \hat{e}_2^{S}) \cdot \mathbf{d}_2 - (\hat{e}_1^{1B} - \hat{e}_1^{S}) \cdot \mathbf{d}_1 = \Delta n_{21}^{1B}\lambda ,$$

$$(\hat{e}_2^{2B} - \hat{e}_2^{S}) \cdot \mathbf{d}_2 - (\hat{e}_1^{2B} - \hat{e}_1^{S}) \cdot \mathbf{d}_1 = \Delta n_{21}^{2B}\lambda ,$$

$$(\hat{e}_2^{3B} - \hat{e}_2^{S}) \cdot \mathbf{d}_2 - (\hat{e}_1^{3B} - \hat{e}_1^{S}) \cdot \mathbf{d}_1 = \Delta n_{21}^{3B}\lambda , \qquad (2.122)$$

$$\overline{\phantom{(\hat{e}_2^{6B} - \hat{e}_2^{S}) \cdot \mathbf{d}_2 - (\hat{e}_1^{6B} - \hat{e}_1^{S}) \cdot}} $$

$$(\hat{e}_2^{6B} - \hat{e}_2^{S}) \cdot \mathbf{d}_2 - (\hat{e}_1^{6B} - \hat{e}_1^{S}) \cdot \mathbf{d}_1 = \Delta n_{21}^{6B}\lambda .$$

By the same reasoning, one can show that when three points Q_1, Q_2, Q_3 are selected on an object, the area of the surface containing the points will have to be observed from five points. obviously, the system of equations for finding $\mathbf{d}_1$, $\mathbf{d}_2$, $\mathbf{d}_3$ will, in this case, be overdetermined with one redundant equation. For four points on the surface, Q_1, Q_2, Q_3 and Q_4, four observation points are needed making the number of unknowns equal to the number of equations. If more than four displacements must be found, the minimum number of observation points will always be four, thus yielding an overdetermined system.

One can readily see that the interferometer arrangement based on the above method of interferogram interpretation is preferentially sensitive to the out-of-plane component of the displacement vector. In addition, the method is obviously cumbersome and time-consuming. However, this method of interpretation can be used to advantage in multi-hologram setups [2.102].

2.3.6 Interferogram Interpretation Based on Fringe Spacing

The spacing between interference fringes can also be employed for the interpretation of holographic interferograms [2.84, 103, 104]. Let the fringe order at the point Q_0 of the object surface be $n(Q_0)$. Then the fringe order at the point Q may be represented by a Taylor expansion

$$n(Q) = n(Q_0) + \frac{\partial n}{\partial x_i}(x_i - x_{i0}) + \frac{1}{2}\frac{\partial^2 n}{\partial x_i^2}(x_i - x_{10})^2 + \dots , \qquad (2.123)$$

where $x_i - x_{i0}$ is the projection of the vector connecting the points Q and Q_0 on the coordinate axis x_i ($i = 1, 2, 3$). Limiting ourselves to the first terms of (2.123) we obtain

$$\Delta n = n(Q) - n(Q_0) = \frac{\partial n}{\partial x_i}(x_i - x_{i0}) . \qquad (2.124)$$

If the points Q and Q_0 are at adjacent fringes and the straight line passing through them is parallel to the coordinate axis x_j, then $\Delta n = 1$ in (2.124), which then transforms to

$$(\partial n/\partial x_j)\Delta x_j = 1 , \qquad (2.125)$$

where Δx_j is the fringe spacing along the x_j axis.

By using the notation with recurrent indices over which summation is carried out, the principal relation of holographic interferometry (2.44) can be rewritten as

$$k_i d_i = n\lambda \quad (i = 1, 2, 3) , \qquad (2.126)$$

where k_i and d_i are the components of the sensitivity and displacement vectors along the ith Cartesian axis. By differentiating (2.126) we arrive at

$$\frac{\partial k_i}{\partial x_j}d_i + k_i\frac{\partial d_i}{\partial x_j} = \frac{\partial n}{\partial x_j}\lambda . \qquad (2.127)$$

Substituting (2.125) into (2.127) yields

$$\frac{\partial k_i}{\partial x_j} d_i + k_i \frac{\partial d_i}{\partial x_j} = \frac{\lambda}{\Delta x_j} . \qquad (2.128)$$

This equation connects the fringe spacing, displacement components and derivatives of the displacement at the point in question.

For a plane object surface lying in the (x_2, x_3) plane, (2.126) can be rewritten

$$\frac{\partial k_1}{\partial x_2} d_1 + k_1 \frac{\partial d_1}{\partial x_2} + \frac{\partial k_2}{\partial x_2} d_2 + k_2 \frac{\partial d_2}{\partial x_2} + \frac{\partial k_3}{\partial x_2} d_3 + k_3 \frac{\partial d_3}{\partial x_2} = \frac{\lambda}{\Delta x_2} , \qquad (2.129)$$

$$\frac{\partial k_1}{\partial x_3} d_1 + k_1 \frac{\partial d_1}{\partial x_3} + \frac{\partial k_2}{\partial x_3} d_2 + k_2 \frac{\partial d_2}{\partial x_3} + \frac{\partial k_3}{\partial x_3} d_3 + k_3 \frac{\partial d_3}{\partial x_3} = \frac{\lambda}{\Delta x_3} , \qquad (2.130)$$

where Δx_2 and Δx_3 are the fringe spacings along the coordinate axes x_2 and x_3, respectively. Similar relations can be constructed for the other two directions of observation. A combined solution of three equations of the type (2.126) and of six equations of the type (2.129,130) yields the components of the displacement vector d_i ($i = 1, 2, 3$) and its derivatives $\partial d_i / \partial x_j$ ($i = 1, 2, 3$; $j = 2, 3$).

2.3.7 Measurement of Displacement by Spatial Filtering of the Object Wave Field

The methods of spatial filtering of the object wave field find broad application in holographic interferometry. We have already discussed the use of spatial filtering in the plane of a doubly exposed hologram illuminated by an unexpanded laser beam to produce high-quality interferograms (Fig. 2.33). The telecentric system of observation is likewise based on the filtering of low spatial frequencies of the object wave field in the Fourier-transform plane. Isolation of strain-induced displacements against the background of in-plane translations of an object can also be achieved by filtering the object wave field in the Fourier transform plane [2.53, 54].

Gates [2.105] was the first to proposed spatial filtering of the object wave field for the interpretation of holographic interferograms. A more appropriate method for displacement measurements is based on a modification described by *Boone* and co-workers [2.106, 107]. It entails isolating a small part of the real image of the object surface by means of a suitable aperture and observing interference fringes in the far-field of Fraunhofer diffraction. The two object wave fields from the first and second exposures recorded before and after the displacement produce a fringe pattern in the Frauenhofer-diffraction field which characterizes the displacement. In this case, the object wave field is filtered in the region of the real image. We now consider this method in more detail.

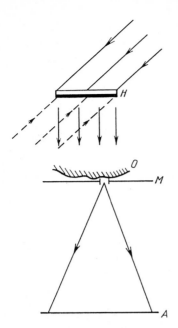

Fig.2.40. Spatial filtration of object field

Figure 2.40 illustrates the object wave filtering in the plane of the real image. A doubly exposed hologram H is illuminated with a wave conjugate to the reference and reconstructs the real image of the object surface O. The dashed lines show the reference wave used in the hologram recording. The mask M with a small aperture isolates the region of the real image and the interface fringes are observed on the screen A.

The two parts of the real image of the object surface isolated by the aperture are equivalent to two monochromatic sources spatially displaced relative to one another. The distance between these sources is equal to the displacement of the given point on the object surface. Thus, in order to determine the fringe pattern on A (Fig.2.40) we have to consider the interference between two point sources Q and Q′ located at the points $(x_1+\Delta x_1,0,0)$ and $(x_1-\Delta x_1,0,0,)$ of the coordinate system shown in Fig. 2.41. The interference pattern is represented by hyperboloids of revolution with the axis x_1 and described by [2.107]

$$x_1^2/a^2 - (x_2^2 + x_3^2)/b^2 = 1 \; ,$$

where $2a = (n+\frac{1}{2})\lambda$ is for the dark fringes. Figure 2.41 shows the intersection of these hyperboloids of revolution with the $x_1 x_2$ plane. We assume that the displacement vector lies in the $x_1 x_2$ plane. The equation of a hyperbola in this plane is

$$(x_1/a)^2 - (x_2/b)^2 = 1 \; . \tag{2.131}$$

At a distance sufficiently far away from the point Q the hyperbolas described by (2.131) can be approximated by their asymptotes:

86

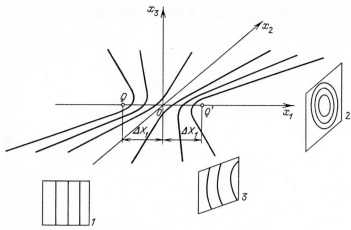

Fig.2.41. Shape of interference fringes for different displacements of a point relative to observer: (*1*) in-plane, (*2*) out-of-plane, and (*3*) intermediate case

$$x_2 = \pm (b/a)x_1 . \qquad (2.132)$$

Thus, with the screen A placed in the position labelled by *1* in Fig.2.41 one will observe Young's fringes describing in-plane displacement of the point. If the screen is in position *2*, the displacement is normal to the object surface and the interference fringes are similar to Newton's rings. With the screen in position *3*, we have a case of intermediate orientation of the displacement vector. The sensitivity of the above method with respect to the out-of-plane component of displacement is substantially lower than to the in-plane one. Therefore, this method is presently used to determine the in--plane component of the displacement vector $2\Delta X_1$

$$2\Delta X_1 = \lambda/\Delta\alpha , \qquad (2.133)$$

where $\Delta\alpha$ is the angular separation between Young's fringes determined from

$$\Delta\alpha = (n-1)L/R .$$

Here n is the number of fringes, L is the fringe spacing, and R is the aperture-to-hologram distance. The magnitudes of out-of-plane displacements can be found from the fringes observed on the virtual image of the object surface. The method of fringe pattern interpretation involving spatial filtering is appropriate to solve problems where the in-plane component of the displacement vector is substantially greater than the out-of-plane one. This method has been used, in particular, to evaluate complex structures [2.108].

3. Optimization of Holographic Interferometers

In this chapter we consider the criteria for designing the optimum holographic interferometer to determine the displacement vector of a point on the surface of a strained body. There are three major approaches presently available to specify or improve the interferometer parameters, that minimize the errors in the displacement vector determination. The most widely used method is simply to compare the holographic results with values derived from other techniques of measurement or obtained analytically or by computation. The degree to which the various results agree serves as a guide to improving the holography apparatus. In this way, for instance, one can estimate the necessary redundancy in the measurements to achieve the desired accuracy in different methods of interferogram interpretation [3.1-5]. One can also decide which parameters should be altered so that the displacement vectors can be determined with the minimum number of equations, namely three [3.6]. It should be stressed that this approach does not yield any information on the lowest possible values of the errors.

The second method is based on a statistical estimate of the errors. Such an approach may serve as a basis for a comparative analysis of optical configurations by computer simulation [3.7-11].

The third method makes use of a nonstatistical analysis of the errors. As discussed in Chap.2, the displacement-vector components are related through a system of linear algebraic equations to the sensitivity matrix. In this method of error estimation, a perturbation is introduced into the right-hand side of the matrix equation for the determination of displacement-vector components and into the sensitivity matrix [3.12-19]. Such an analysis provides a priori information on the precision with which the displacement vector and its components can be determined.

The present chapter deals with the latter two methods, based on statistical and nonstatistical approaches. We limit ourselves to the two major techniques of holographic interferogram interpretation, namely, the absolute and relative fringe-order determination. The main principles underlying the construction of interferometer setups, based on the use of the so-called criterion of C-optimality, are discussed for the cases where off-axis and opposed beam holograms are used to measure two or three components of the displacement vector. A comparison is made of the methods of interferogram interpretation involving absolute and relative fringe-order readout. An original technique which uses overlay holographic interferometers, with reflection holograms to measure the in-plane components of a displacement vector, is presented.

3.1 Errors in Measurements of Displacement

3.1.1 Statistical Analysis of Errors

Most of the methods for holographic-interferogram interpretation discussed in the preceding chapter and employed in practice to measure the three components of the displacement vector $\mathbf{X}$, involve the solution of a system of linear equations of the type

$$\mathbf{KX} = \lambda\mathbf{Z} , \qquad (3.1)$$

where $\mathbf{K}$ is the 3×3 sensitivity matrix characterizing the specific optical configuration of the holographic interferometer used, $\mathbf{Z}$ is a vector describing the interference pattern, and λ is the wavelength of light. For various methods of interpretation these systems of equations differ in type and dimension of the sensitivity matrix $\mathbf{K}$, as well as in the way in which the vector $\mathbf{Z}$ is specified. What all cases have in common is that $\mathbf{K}$ is given by the experimental conditions while the vector $\mathbf{Z}$ is determined from interference fringes, and both the matrix elements and the vector components are known only to within certain errors. We first analyze the effect of each source of error on the accuracy of the displacement-vector components by a statistical approach [3.3-11].

Consider the case where the sensitivity matrix is precisely known, whereas the fringe orders are counted to within an error given by the vector $\Delta\mathbf{Z}$, which, for the absolute fringe orders and the minimum required number of equations (three), can be written as

$$\Delta\mathbf{Z} = [\Delta n_1, \Delta n_2, \Delta n_3]^{\mathrm{T}} .$$

As follows from (3.1), the displacement-error vector $\Delta\mathbf{X}$ is related to the fringe-order error vector through the expression

$$\Delta\mathbf{X} = \lambda\mathbf{K}^{-1}\Delta\mathbf{Z} , \qquad (3.2)$$

where $\mathbf{K}^{-1}$ is the inverse matrix.

We now assume that the measurements of the fringe-order number at a given point and for a fixed viewing direction i are repeated a sufficiently large number of times so that the error Δn_i can be considered a random variable obeying the normal-distribution law with a zero mean. We also assume that the errors Δn_i are independent for different viewing directions. Then we may write the covariance matrix Γ for the error vector $\Delta\mathbf{Z}$ in the form:

$$\Gamma = \begin{bmatrix} \sigma_{n1} & 0 & 0 \\ 0 & \sigma_{n2} & 0 \\ 0 & 0 & \sigma_{n3} \end{bmatrix} , \qquad (3.3)$$

where σ_{ni} (i = 1, 2, 3) is the standard deviation for fringe-order measurement in the chosen viewing directions. If the standard deviations for all the viewing directions are equal

$$\sigma_{n1} = \sigma_{n2} = \sigma_{n3} = \sigma \,, \tag{3.4}$$

then for the covariance matrix we have

$$\Gamma = \sigma I \,, \tag{3.5}$$

where I is the unitary matrix.

The covariance matrix L for the displacement vector can be written as [3.8]

$$L = \lambda K^{-1}\Gamma(K^{-1})^T \,. \tag{3.6}$$

Substituting (3.5) into (3.6) yields

$$L = \lambda^2\sigma(K^TK)^{-1} \,. \tag{3.7}$$

The square root of the diagonal elements of the matrix $(K^TK)^{-1}$ gives us the measurement errors for the components X_1, X_2 and X_3 of the displacement vector in the form σ_{X1}/σ, σ_{X2}/σ and σ_{X3}/σ, respectively. Thus, for the expression for the measurement errors of the displacement vector with known both interferometer sensitivity matrix and fringe order measurement variance we obtain

$$\sigma_{xi}^2/\sigma^2 = \lambda(K^TK)^{-1}_{ii} \quad (i = 1, 2, 3) \,. \tag{3.8}$$

We now consider the fringe-order vector to be accurately known, and the sensitivity matrix elements to be determined with an error given by the matrix ΔK. As follows from the principal equation (3.1), the displacement ΔX is related to the error matrix:

$$\Delta X = X K^{-1}\Delta K \,. \tag{3.9}$$

We now assume that the errors of the sensitivity matrix elements are random variables distributed normally with a zero mean and a variance σ_K^2. Following a similar reasoning as above and taking the elements of the ΔK matrix to be uncorrelated, one can obtain, using (3.9), an expression to estimate the errors of the displacement-vector components due to inaccurate measurements of the sensitivity matrix elements:

$$\sigma_{xi}^2/\sigma_K^2 = X^2(K^TK)^{-1}_{ii} \quad (i = 1, 2, 3) \,. \tag{3.10}$$

The square roots of the diagonal elements of $(K^TK)^{-1}$ yield the expected errors of the displacement-vector components in the form $\sigma_{x1}/\sigma_K X$, $\sigma_{x2}/\sigma_K X$, and $\sigma_{x3}/\sigma_K X$.

90

Nobis and *Vest* [3.8] showed that the major contribution to the uncertainty of displacement vector determination comes from errors in counting the fringe order. They suggested, similar to the conclusion of *Dhir* and *Sikora* [3.1], to use an overdetermined system of equations as (3.1). In a computer simulation based on relative fringe order, *Ek* and *Biedermann* [3.20] established that the errors of the in-plane components grow linearly with increasing distance from the object whereas that of the normal component has a quadratic increase. Another statistical approach has been proposed by *Lisin* [3.21].

It should be noted that the approach of error estimation discussed in this section allows a comparison of optical interferometers. However, it cannot provide a guide in selecting the optical system ensuring the lowest possible error in displacement measurement.

3.1.2 Nonstatistical Analysis of Displacement Measurement Error

The presence in (3.1) of an a-priori known sensitivity matrix K as well as the linear relation between the measured (Z) and unknown (X) parameters permits application of a nonstatistical approach to the evaluation of the displacement error. We express the error in the vector Z also in the form of a vector ΔZ, and that in the elements of the matrix K, by a matrix ΔK. The components ΔZ_j (j = 1, 2, 3) represent the errors made in measuring the positions of the interference fringes. The elements of the matrix ΔK are related to the errors of the directly measured linear or angular parameters of the interferometer system through the following expressions [3.22]

$$\Delta K_{ij} = \sum_{m=1}^{\ell} \frac{\partial k_{ij}}{\partial \theta_m} d\theta_m , \qquad (3.11)$$

where θ_m denotes the generalized parameters of the optical system entering the elements of the matrix K; $d\theta_m$ is the measurement error for the mth parameter, and ℓ is the number of the generalized parameters.

Perturbations present in the original data lead to an uncertainty in the solution of (3.1). The upper bound on the error made in calculating the displacement vector ΔX can be estimated in the following way [3.23, 24]

$$\frac{||\Delta X||}{||X||} \leq \frac{||K|| \cdot ||K^{-1}||}{1 - ||K|| \cdot ||K^{-1}|| (||\Delta K||/||K||)} \left(\frac{||\Delta Z||}{||Z||} + \frac{||\Delta K||}{||K||} \right) , \qquad (3.12)$$

where ||...|| are the vector and matrix norms compatible with one another.

The quantity $||K|| \cdot ||K^{-1}||$ is denoted cond{K} and called the *condition number of a system of linear algebraic equations*.

Since the condition number of a system depends only on the actual kind of the sensitivity matrix K, i.e., on the geometry of an interferometer's optical system, (3.12) permits a preliminary evaluation of the errors in the absolute value of the displacement vector. The errors in the individual components of X can be estimated by means of the inequality [3.25]

$$|\Delta X_i| \le \gamma_p \sum_{j=1}^{3} |K_{ij}^{-1}| \quad (i = 1, 2, 3) , \tag{3.13}$$

where $\gamma_p = \beta_Z + \beta_K \Sigma_{i=1}^3 |X_i|$; β_Z and β_K are the maximum elements of the vector ΔZ and of the matrix ΔK, respectively.

Inequality (3.13) relates the absolute errors ΔX_i and allows estimation of the accuracy of the individual components X_i. By using (3.13), one can express the norm $\|\Delta X\|$ in terms of any component of the vector, e.g.,

$$|\Delta X_i| = |\Delta X_1| G_i , \quad \|\Delta X\| = |\Delta X_1| \cdot \|G\| , \tag{3.14}$$

where

$$G_i = \left(\sum_{j=1}^{3} |K_{ij}^{-1}| \right) \left(\sum_{j=1}^{3} K_{1j}^{-1} \right)^{-1} \quad (i = 1, 2, 3) .$$

Combining (3.14 and 12) yields the desired estimate for the errors on the displacement components [3.23]

$$|\Delta X_i| \le \frac{\text{cond}\{K\}}{1 - \text{cond}\{K\} \cdot \left(\frac{\|\Delta K\|}{\|K\|} \right)} \frac{G_i}{\|G\|} \left(\frac{\|\Delta Z\|}{\|Z\|} + \frac{\|\Delta K\|}{\|K\|} \right) . \tag{3.15}$$

In actual experiments the criterion governing the quality of a holographic measurement is the accuracy with which all three displacement components can be determined. The desired accuracy can be attained by properly selecting the coefficients of the coupled equations (3.1), i.e., of the parameters of the optical interferometer. As seen from estimates, (3.12 and 15), the nonstatistical analysis when assessing various experimental designs consists in choosing a system of equations (3.1) with a minimum condition number. The condition number characterizes the degree of sensitivity of a system of linear algebraic equations to a perturbation on the quantities K and Z. The optimum criteria, in accordance to which one chooses the matrix K of the sensitivity of a holographic interferometer with the minimum condition number, is called the C-optimality criteria.

The quantity cond$\{K\}$ depends on the actual type of the matrix norm used. In practice, spectral condition numbers are most convenient since linear transformations can be carried out with orthogonal matrices. From the definition of the spectral condition number it follows [3.24] that

$$\text{cond}\{K\} = \|K\| \cdot \|K^{-1}\| = \frac{\mu_1}{\mu_3} \ge 1 , \tag{3.16}$$

where μ_1 and μ_3 are the maximum and minimum singular numbers of the matrix K which are equal to the maximum and minimum square roots of the eigenvalues of $(KK^T)^{-1}$.

As follows from (3.16), the theoretically possible minimum condition number is unity. The experimental designs meeting this condition are C optimal. Thus, finding a design satisfying C-optimality consists in finding a sensitivity matrix with the minimum condition number:

$$\text{cond}\{K\} = 1 . \tag{3.17}$$

The necessary and sufficient condition for this is the orthogonality of the vectors constructed from the columns (rows) of the matrix K and the equality of their dimensions. The matrices possessing these properties are called quasi-orthogonal. When condition (3.17) is met, all components of the displacement vector X are equally accurate and have the minimum error which can be estimated with the expression

$$|\Delta X_i| \leq \frac{1}{\sqrt{3}(1 - \|\Delta K\|/\|K\|)} \left(\frac{\|\Delta Z\|}{\|Z\|} + \frac{\|\Delta K\|}{\|K\|} \right) . \tag{3.18}$$

The next two subsections discuss the application of the principles of nonstatistical design to the methods of interferogram interpretation based on the relative and absolute fringe order.

3.2 Design of a Holographic Experiment

3.2.1 Selection of Interferometer Parameters for Relative Fringe-Order Readout

The method of interferogram interpretation from relative fringe orders used to determine surface displacements of a strained object is attractive primarily because it does not require identification of the zero-motion fringe. The displacement-vector components at a given point on the object surface are found by solving the system of (2.114) with the following form

$$KX = \lambda \delta N , \tag{3.19}$$

where K is a 3×3 sensitivity matrix whose elements are the projections of the sensitivity vectors k_{21}, k_{43}, k_{65} onto the coordinate axes x_i ($i = 1, 2, 3$), see Sect.2.3.4; $\delta N = [n_{21}, n_{43}, n_{65}]^T$ is the relative fringe-order vector whose components are equal to the number of the fringes that have passed through the given point on the object surface as the viewing direction is varied.

In practice, the method of interpretation based on relative fringe orders is complicated by instability of the measurement. Therefore, it is recommended that enough viewing directions within the aperture of one hologram be chosen so that an overdetermined system of at least six equations can be generated [3.3]. Another approach is to use a double-hologram interferometer configuration to generate an overdetermined system of ten equations [3.2].

The criterion of C-optimality discussed in Sect.3.1.2 can also be used to select viewing directions. Computer simulation [3.26] has been adopted to find optimum systems within the constraints imposed on the interferometer parameters. The displacements obtained in this way far exceed the minimum values derived from theory, however. By lifting the constraints on the allowed range of variation of the viewing directions and by using only a diagonal or symmetrical matrix K, one obtains a trivial optical system with orthogonal sensitivity vectors requiring construction of three holograms [3.15].

By satisfying the optimality condition (3.17) one can minimize the errors on the displacement-vector components according to (3.18). To follow such a procedure, one has to select quasi-orthogonal matrices with elements corresponding to the interferometer sensitivity matrix.

As an illustration, take for example a matrix T_1 derived from a canonical orthogonal matrix

$$T_1 = A_1 \begin{bmatrix} 1/\sqrt{2} & -1/\sqrt{2} & 0 \\ 1/\sqrt{2} & 1/\sqrt{2} & 0 \\ 0 & 0 & 1 \end{bmatrix}, \tag{3.20}$$

and a matrix T_2

$$T_2 = A_2 \begin{bmatrix} \sqrt{3}/3 & \sqrt{6}/3 & 0 \\ \sqrt{3}/3 & -\sqrt{6}/6 & -\sqrt{2}/\sqrt{2} \\ \sqrt{3}/3 & -\sqrt{6}/6 & \sqrt{2}/2 \end{bmatrix}. \tag{3.21}$$

As seen, a necessary condition for the construction of an experimental design is a proper distribution of the signs of the sensitivity matrix elements. Figure 3.1 shows two positions of the photographic plate relative to the origin 0 (the point under study) and eight observation points providing all possible sign combinations in the rows of the sensitivity matrix K. To make the analysis more visual, the holograms H_1 and H_2 are assumed to be perpendicular to the lines connecting their centers and the point 0. The lines $h_1 h_1$ and $h_2 h_2$ in the $x_1 x_2$ plane divide the holograms in half. The x_1 axis coincides with the normal to the object surface at point 0. The distances ℓ from the hologram centers to the observation points on the holograms are equal. Under these conditions the sensitivity vectors $k_{pq} = \hat{e}_p - \hat{e}_q$ $(p,q = 1,2,3,4; \; p \neq q)$ and $k_{rt} = \hat{e}_r - \hat{e}_t$ $(r,t = 5,6,7,8; \; r \neq t)$ lie in planes parallel to those of the holograms H_1 and H_2., respectively. For the sake of simplicity, they are indicated to lie in the planes of the holograms (Fig.3.1). The interferometer-sensitivity matrix is formed by rows of the following kind:

94

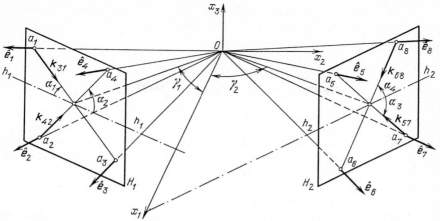

Fig.3.1. Two-hologram interferometer setup for fringe interpretation based on relative fringe order counting

$$[k_{j1}, k_{j2}, k_{j3}] = \frac{2\ell}{\sqrt{\ell^2 + R^2}} (\pm\cos\alpha_j \sin\gamma_m, \pm\cos\alpha_j \cos\gamma_m, \pm\sin\alpha_j) , \qquad (3.22)$$

where $j = 1,2,...,n$ is the number of transitions on the hologram, m is the hologram number, and R is the distance from the hologram centers to the point 0.

The sign combinations of the elements k_{ij} and the corresponding sensitivity vectors are listed in Table 3.1. To construct an interferometer based on T_1, one can choose two transitions of the observation points on the hologram H_1 yielding the sensitivity vectors k_{31} and k_{42}, and on the hologram H_2, a transition which will provide the sensitivity vector k_{68}. Then for the sensitivity matrix we obtain

$$k_{T_1} = \frac{2\ell}{\sqrt{\ell^2 + R^2}} \begin{bmatrix} \cos\alpha_4 \sin\gamma_2 & -\cos\alpha_4 \cos\gamma_2 & -\sin\alpha_4 \\ \cos\alpha_1 \sin\gamma_1 & \cos\alpha_1 \cos\gamma_1 & -\sin\alpha_1 \\ \cos\alpha_2 \sin\gamma_1 & \cos\alpha_2 \cos\gamma_2 & \sin\alpha_2 \end{bmatrix} . \qquad (3.23)$$

Table 3.1. Signs of sensitivity-matrix row elements

Matrix element	Sensitivity vector							
	Hologram H_1				Hologram H_2			
	$\hat{e}_4$-$\hat{e}_2$	$\hat{e}_2$-$\hat{e}_4$	$\hat{e}_3$-$\hat{e}_1$	$\hat{e}_1$-$\hat{e}_3$	$\hat{e}_8$-$\hat{e}_6$	$\hat{e}_6$-$\hat{e}_8$	$\hat{e}_5$-$\hat{e}_7$	$\hat{e}_7$-$\hat{e}_5$
K_{j1}	+	−	+	−	−	+	+	−
K_{j2}	+	−	+	−	+	−	−	+
K_{j3}	+	−	−	+	+	−	+	−

95

If we select the sensitivity vectors k_{31} on the hologram H_1, and k_{57}, k_{68} on the hologram H_2, we arrive at an interferometer with a sign distribution corresponding to the matrix T_2

$$K_{T_2} = \frac{2\ell}{\sqrt{\ell^2+R^2}} \begin{bmatrix} \cos\alpha_1\sin\gamma_1 & \cos\alpha_1\cos\gamma_1 & -\sin\alpha_1 \\ \cos\alpha_4\sin\alpha_4 & -\cos\alpha_4\cos\gamma_2 & -\sin\alpha_4 \\ \cos\alpha_3\sin\gamma_2 & -\cos\alpha_3\cos\gamma_2 & \sin\alpha_3 \end{bmatrix} . \qquad (3.24)$$

It should be stressed that a variation of the viewing direction within the aperture of only one hologram H_1 or H_2 cannot yield a sign distribution of the elements k_{ij} which would coincide with that of T_1, (3.20), or T_2, (3.21). By using a proper combination of observation point transitions on the two holograms and requiring that the elements k_{ij} in the form (3.23 and 24) be equal to the corresponding elements of the matrices (3.20 and 21), one can determine the parameters of the interferometer for which the relations (3.17 and 18) would be met. Values of these parameters for the matrix (3.23)

$$\alpha_1 = \alpha_4 = 0 , \quad \alpha_2 = 90° , \quad \gamma_1 = \gamma_2 = 45° , \quad A_1 = 2\ell(\ell^2+R^2)^{-1/2} , \qquad (3.25)$$

and for the matrix (3.24) are

$$\alpha_1 = 0 , \quad \alpha_3 = \alpha_4 = 45° , \quad \gamma_1 = 35,3° , \quad \gamma_2 = 54,7° , \quad A_2 = 2\ell(\ell^2+R^2)^{-1/2} . \qquad (3.26)$$

Thus the two-hologram interferometer setups with parameters satisfying (3.25 or 26) provide equally accurate measurements of the displacement vector components X_i ($i = 1,2,3$) for a point 0 with a maximum accuracy, as seen from (3.18), compatible with the errors $\Delta(\delta N)$ and ΔK.

Besides the quasi-orthogonal matrices of the smallest possible dimensions 3×3 given above, one can also use a 4×3 matrix composed of any three columns of the matrix

$$Q = A_3 \begin{bmatrix} 1 & 1 & 1 & -1 \\ 1 & 1 & -1 & 1 \\ 1 & -1 & 1 & 1 \\ -1 & 1 & 1 & 1 \end{bmatrix} . \qquad (3.27)$$

To estimate the errors by solving (3.18), one then uses the matrix K^+ in place of K:

$$K^+ = (K^T K)^{-1} K^T .$$

The optimum experimental design corresponding to the first three columns of (3.27) and the sensitivity vectors k_{42}, k_{31}, k_{57} and k_{86} (Fig.3.1) is described by the following parameters:

$$\alpha_1 = \alpha_2 = \alpha_3 = 35,3° , \quad \gamma_1 = \gamma_2 = 45° , \quad A_3 = 2\sqrt{3}\ell(\ell^2+R^2)^{-1/2} . \qquad (3.28)$$

Note that in all the systems thus obtained, the angle between the lines connecting the point 0 with the centers of the photographic plates is 90°. The selection of an optimum setup for the viewing directions and of the hologram center-to-origin distances is usually governed by the desire to reduce the error on the vector δN (3.19). The decisive factor will, in most cases, be the number of fringes which have passed through the point in question on the object surface as the viewing direction is varied. Therefore it is wise to consider the expediency of using overdetermined systems of the type (3.28). Starting from (3.10,22) and the known properties of matrix norms, it can be shown that the relative error $\|\Delta K\| \cdot (\|K\|)^{-1}$ is the same for the sensitivity matrices of the parameters (3.25, 26 or 28). Hence, according to the inequality in (3.18), for equal relative errors in the individual measurements $\Delta(\delta N_j) \cdot (\delta N_j)^{-1}$ (j = 1, 2, 3, ...n, n = 3 or 4) corresponding to one change in the viewing direction, the accuracy of determination of the displacement components for the minimum required matrix dimension of 3×3 and for the greater dimension of 4×3 will be the same. In other words, increasing the number of changes in the viewing direction above three is not necessary, since a further increase in the volume of measurements does not bring about any improvement in accuracy.

The accuracy gained as a result of optimizing the interferometer might be demonstrated best by a comparison of the one- and two-hologram arrangement. We begin with the worst possible scenario: we consider an interferometer described by the parameters (3.26) but with a single hologram H_2 set at an angle $\gamma_2 = 54.7°$ to the x_1-axis. We represent the change in the viewing direction by the hologram H_1 with $\alpha_1 = 0$ (Fig. 3.1). This operation is equivalent to replacing the first row in the matrix T_2 by $(\pm\sqrt{6}/3, \pm\sqrt{3}/3, 0)$. One can readily see that the matrix thus constructed is singular (cond{K} $\rightarrow \infty$) which makes displacement measurements with a one-hologram interferometer, described by such a matrix, meaningless. Moreover, for the changes in the viewing direction on one hologram considered for this case, the relation cond{K}$\rightarrow\infty$ is valid for any angle γ from 0 to 90°. This accounts for the frequently observed instability in the measurement of displacement-vector components from the relative fringe orders on one-hologram. In addition, it is usually aggravated by a large relative error in the determination of the vector δN.

The accuracy limits of a one-hologram interferometer can be derived from a computer-simulated optimization [3.26]. A numerical simulation carried out for a 160×100 mm^2 photographic plate located 300 mm away from the point of interest showed that the minimum condition number of the sensitivity matrix is as high as fifteen! Therefore, although the condition number defines the maximum error limit it appears that two-hologram optical systems provide an improvement in the accuracy in the determination of displacement-vector components of at least an order of magnitude.

One must also consider the deviations of the geometrical parameters from the optimum values, which are beyond reach in studies of extended objects. They arise from the fact that it is impossible to obtain a sensitivity matrix which is constant for all points on a large surface. A computed dependence of the magnitude of cond{K} on the distance x_2 from point 0 on a

97

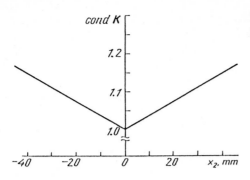

Fig.3.2. Condition number vs. distance from the origin

plane surface for which cond{K} = 1 is presented in Fig.3.2 [3.13]. The distance from the origin to the hologram centers is R = 0.3 m, the distance between the observation points ℓ is 11 cm. One can see that within the range x_2 = ±50 mm, the values of cond{K} differ from unity by not more than 0.2, i.e., they are practically invariant.

3.2.2 Interferometer Design for Absolute Fringe-Order Readout

Interferogram interpretation based on absolute fringe orders is the most efficient method to determine displacement vectors of points on surfaces of an arbitrary shape, provided that their directions are unknown. In this case the unknown displacement vector **X** is derived from the solution of (3.1) which takes on the form

$$\mathbf{K X} = \lambda \mathbf{N} \ , \tag{3.29}$$

where K is a 3×3 sensitivity matrix whose elements are equal to the projections of the difference between the unit vectors of observation and illumination on the coordinate axes x_i (i = 1, 2, 3); $\mathbf{N} = [n_1 \ n_2 \ n_3]^T$ is the absolute fringe-order vector.

When absolute fringe orders are used to interpret an interferogram, the interferogram sensitivity with respect to the components normal to the surface of the object is known to exceed, sometimes substantially, that for the in-plane components. As a result of this anisotropy in the sensitivity, it is impossible to construct an optimum design (cond{K} = 1). Error estimates made with the inequalities (3.15, 18) become very conservative, and, hence, cannot be recommended for practical use.

In searching for ways to estimate minimum errors in such situations one can use the dependence of the condition number on the scale of the right-hand side of (3.29). It should be noted that in this method of fringe interpretation, the part of the relative error coming from the optical system which is due to measurements of the geometrical parameters (i.e., of the sensitivity matrix ||ΔK||/||K||) is small enough to be negligible [3.14]. Then inequality (3.15) which is used to estimate the errors on the displacement-vector components takes on the form

$$\frac{|\Delta X_i|}{\|X\|} \le \text{cond}\{K\} \frac{G_i}{\|G\|} \frac{\|\Delta N\|}{\|N\|} , \qquad (3.30)$$

where ΔN is the error in the determination of N.

Transforming the variables in (3.29), $X = D_2 X^*$ in (3.29) yields an equivalent system which can be more conveniently used to analyze the measurement accuracy .

$$K^* X^* = \lambda N , \qquad (3.31)$$

where $K^* = KD_2$ and $X^* = D_2^{-1}X$ are the scaled sensitivity matrix and displacement vector, respectively, and D_2 is a 3×3 diagonal matrix. For the optimum condition which provides minimization of the errors made in measurements of the displacement components with an interferometer specified by (3.31) we obtain

$$\text{cond}\{K^*\} = 1 . \qquad (3.32)$$

The inequality (3.30) allows the derivation of an expression for estimating the errors on the components ΔX_i^* for the case of a quasi-orthogonal matrix K^* which satisfies condition (3.32)

$$|\Delta X_i^*|/\|X^*\| \le \|\Delta N\|/\sqrt{3}\|N\| \quad (i = 1,2,3) . \qquad (3.33)$$

The components of the displacement vector X and their errors reduced to the original scale can be determined from

$$(X_i \pm \Delta X_i) = D_2(i) (X_i^* \pm \Delta X_i^*) , \qquad (3.34)$$

where $D_2(i)$ are the elements of the matrix D_2. The construction of optimum interferometer setups satisfying the estimates (3.33) thus requires choosing a matrix K in the original scale for which a matrix D_2 satisfying condition (3.32) exists.

To illustrate the solution of the optimization problem with the above algorithm, we consider an interferometer setup with three viewing directions $\hat{e}_j$ ($j = 1,2,3$) and one direction of illumination, $\hat{e}_s$, coinciding with the x_1 axis (Fig. 3.3). The x_1 axis is directed along the normal to the surface under study at the point 0 serving as the origin of a Cartesian coordinate system x_i ($i = 1,2,3$). The sensitivity matrix K at the point 0 has the form

$$K = \begin{bmatrix} 1+\cos\psi_1 & -\sin\psi_1\cos\alpha_1 & \sin\psi_1\sin\alpha_1 \\ 1+\cos\psi_2 & \sin\psi_2\cos\alpha_2 & -\sin\psi_2\sin\alpha_2 \\ 1+\cos\psi_3 & \sin\psi_3\cos\alpha_3 & \sin\psi_3\sin\alpha_3 \end{bmatrix} . \qquad (3.35)$$

The elements of the first column are always greater than unity, and those of the second and third columns less than unity. This represents a mathematical consequence of the fact that when interpreting fringe patterns based on

99

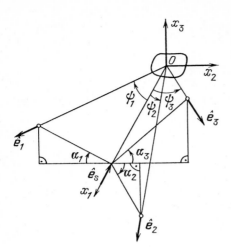

Fig.3.3. Illustration of an experimental design using fringe interpretation based on absolute fringe order counting

absolute orders, holographic interferometers exhibit a higher sensitivity with respect to the out-of-plane component of the displacement vector than to its in-plane components. It follows that the optimality condition (3.17) cannot be met since one of the conditions for the orthogonality of a matrix is that the vectors whose components are elements of the matrix columns should have equal length. Therefore, in order to make the length of the column vectors of K equal, one must apply the scaling matrix D_2

$$D_2 = \begin{bmatrix} 1/A & 0 & 0 \\ 0 & 1/B & 0 \\ 0 & 0 & 1/B \end{bmatrix}, \tag{3.36}$$

where A and B are some coefficients yet to be determined. The matrices K and K^* are related through

$$K = D_2^{-1} K^* . \tag{3.37}$$

We select as the initial quasi-orthogonal matrix K^* a matrix of the form

$$K^* = A_0 \begin{bmatrix} 1 & -1 & \tfrac{1}{2} \\ 1 & \tfrac{1}{2} & -1 \\ \tfrac{1}{2} & 1 & 1 \end{bmatrix}, \tag{3.38}$$

where A_0 is an arbitrary constant.

Substituting (3.36, 38) into (3.37) yields

$$K = A_0 \begin{bmatrix} A & -B & \tfrac{1}{2}B \\ A & \tfrac{1}{2}B & -B \\ \tfrac{1}{2}A & B & B \end{bmatrix} . \tag{3.39}$$

100

A term-by-term comparison of the matrices (3.39 and 35) allows one to express the coefficients A and B in terms of the interferometer parameters. The first two rows of the matrices (3.39 and 35) coincide if

$$\psi_1 = \psi_2 , \quad \alpha_1 = 90° - \alpha_2 , \quad \tan\alpha_1 = \tfrac{1}{2} \quad (\alpha_1 = 26.56°) ,$$
$$A = 1 + \cos\psi_1 , \quad B = \sin\psi_1 \cos\alpha_1 . \tag{3.40}$$

By comparing the elements k_{23} and k_{33} of the matrices (3.35, 39) one finds the angle α_3

$$\cos\alpha_3 = \sin\alpha_3 , \quad \alpha_3 = 45° . \tag{3.41}$$

To construct an optimum design, one must only find the relation between the angles ψ_1 and ψ_3 and establish the range of their variation. By requiring the elements k_{13} and k_{23} of (3.39 and 35) to be equal, we get a system of equations

$$1 + \cos\psi_3 = \tfrac{1}{2}(1 + \cos\psi_1) ,$$
$$\sin\psi_3 = \sin\psi_1 \cos\alpha_1 . \tag{3.42}$$

Solving this system yields the relation between the angles ψ_1 and ψ_3

$$\tan(\psi_3/2) = 2.53 \tan(\psi_1/2) . \tag{3.43}$$

To permit observation in all directions from the point 0 on the surface, the viewing direction vectors $\hat{e}_j$ (j = 1, 2, 3) should be confined within a solid angle of $\pi/2$, implying that $\psi_1 < 90°$ and $\psi_3 < 90°$. Combining these conditions with (3.43), we obtain the ranges for the viewing angles, namely

$$0 < \psi_1 < 43.14° , \quad 0 < \psi_3 < 90° . \tag{3.44}$$

Thus (3.40, 43 and 44) describe a C-optimal experimental design for the determination of the components of the displacement vector for the point 0, constructed on the basis of an orthogonal matrix of the form (3.38). In a similar way, one can obtain other optical configurations by using other quasi-orthogonal matrices with elements relating to the sensitivity of real interferometers.

The errors made in determining the displacement-vector components in the original dimension can be found by substituting into (3.34) the elements of the matrix

$$\Delta \mathbf{X} = \begin{bmatrix} \Delta X_1 \\ \Delta X_2 \\ \Delta X_3 \end{bmatrix} = \begin{bmatrix} (1+\cos\psi_1)^{-1} \\ (\sin\psi_1 \cos\alpha_1)^{-1} \\ (\sin\psi_1 \cos\alpha_1)^{-1} \end{bmatrix} |\Delta \mathbf{X}^*| \tag{3.45}$$

where the value $|\Delta \mathbf{X}^*| \equiv |\Delta X_i^*|$ (i = 1, 2, 3) is found from (3.33). As seen from (3.45), the anisotropy in the interferometer sensitivity results in an error on the normal component ΔX_1, which is always smaller than those of the in-plane components ΔX_2 and ΔX_3.

Meeting the optimality condition $\text{cond}\{K^*\} = 1$ means that at fixed viewing angles ψ_1 and ψ_2, the interferometer is capable of measuring the displacement vector components with minimum errors, as defined by the C criterion. The actual selection of the angles ψ_1 and ψ_3, (3.42), within the allowable range (3.44) is governed, as a rule, by the desire to reach the maximum possible sensitivity of measurement for the in-plane displacement components. In practice, however, the freedom in this selection is limited by the requirement that the part of the object under study be simultaneously observable from three directions. On curved surfaces, the maximum possible sensitivity to in-plane components of the displacement vectors decreases and depends on the actual shape of the object surface and its distance from the observation points. The technical capabilities of a holographic setup, the dimensions of the optical elements, the design features of the loading devices etc. may also contribute to a reduction of the allowed range of ψ_1 and ψ_3, (3.44). The use of collimated illumination and collimated observation keeps the sensitivity vector constant over the entire object surface. In this case (3.32) is satisfied for all points.

In the general case of a spherical illumination wavefront and observation from fixed points, the sensitivity vector k_j ($j = 1, 2, 3$) varies as one moves from one surface point to another. For such optical systems a C-optimal design of the form (3.32) can be constructed for some characteristic point, such as the geometric center of the surface section under study, the center of symmetry of the object, the reference point where displacements must be determined with maximum accuracy, etc. In all these cases, the optimality condition will be satisfied in the vicinity of the point within the constraints imposed on the parameters of the interferometer setup. According to (3.30, 34) the component measurement errors near this point can be estimated by the expression

$$|\Delta X_i| \leq \sqrt{3}\,\Delta n D_2(i)\,\text{cond}\{K^*\}\,\frac{G_i^*}{\|G^*\|}\,\frac{\|X^*\|}{\|N\|} \quad (i = 1, 2, 3) , \qquad (3.46)$$

where Δn is the measurement error on the absolute fringe orders which is assumed to be equal for all viewing directions, $D_2(i)$ is determined from (3.45) for the origin at 0, and G^* is a vector with the components

$$G_i^* = \sum_{j=1}^{3} |(K_{ij}^*)^{-1}| \left[\sum_{j=1}^{3} |(K_{ij}^*)^{-1}| \right]^{-1} \quad (i = 1, 2, 3) .$$

The nonstatistical approach of error analysis described above can be used for a preliminary evaluation of the errors and their minimization through an appropriate choice of the interferometer parameters. This technique has proven to be efficient in solving problems of the mechanics of a strained object, for which no theoretical solution has been found and computational

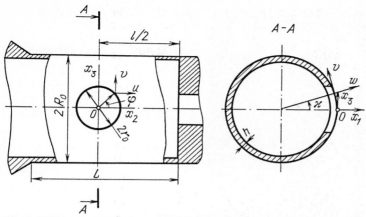

Fig.3.4. Side-on and cross-sectional view of a cylindrical shell with a circular hole; u, v and w are the displacement components of a point at the edge of the hole

methods are inadequately developed. An example of such a problem is the deformation of shell surfaces with holes under different loading regimes [3.27].

The theory for constructing an interferometer system by using the C-optimality criterion with ψ_1, ψ_2, ψ_3 and $\alpha_1, \alpha_2, \alpha_3$ given by (3.40, 43 and 44) can be illustrated by a study of edge deformation of a circular hole in a cylindrical shell subjected to a tensile stress (Fig.3.4). The outer radius of the shell is R = 30 mm, the hole radius r = 12 mm, the length of the cylindrical part L = 100 mm. The shell is made of D16T alloy. A three-hologram interferometer constructed by using the C-optimality criterion for observing the central point of the hole with the parameters ψ_1 = 25° and ψ_3 = 58.6° allows one to view more than one half of the rim of the hole from all three directions. Because the cylinder is symmetrical with respect to two planes passing through the center of the hole perpendicular to its plane, information on the displacements over this small area is sufficient to analyze the deformed state in the vicinity of the hole. A point illumination source is mounted on the x_1 axis 0.83 m away from the shell surface. The distance from the observation points along the vectors $\hat{e}_1$ and $\hat{e}_2$ (Fig.3.3) to the origin is 0.4 m and in the direction of $\hat{e}_3$ it is 0.3 m. The condition number of the scaled sensitivity matrix for the hole edge does not exceed 1.21. Figures 3.5a-c present, respectively, two-exposure interferograms obtained for three viewing directions specified by the unit vectors $\hat{e}_1$, $\hat{e}_2$ and $\hat{e}_3$ (Fig. 3.3), with the loading incremented between the exposures by ΔP = 1.18 kN. As shown by an analysis of the sources and sinks of the fringes, which come forth as the viewing direction is varied within the hologram aperture, the values of the absolute fringe orders on the edge pass through zero four times. The absolute fringe orders at the grid corners along the edge were determined in two ways, namely, by means of an elastic string and by using the zero fringe on the unstrained part of the cylinder. Fractional fringe orders were found by linear approximation. The solution of (3.29) permitted

103

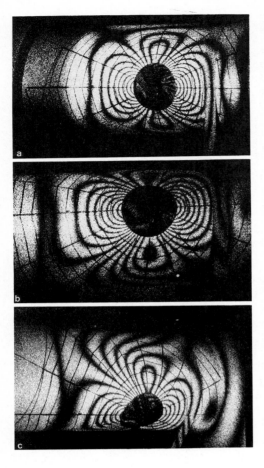

Fig.3.5. Reconstructed image of a cylindrical shell with a circular hole under tension obtained in (a) $\hat{e}_1$, (b) $\hat{e}_2$, and (c) $\hat{e}_3$ directions, see Fig.3.3

determination of the displacement-vector components X_1, X_2 and X_3 in the Cartesian coordinate system chosen. The displacement components u, v, and w in a cylindrical system (Fig.3.4) were derived from

$$u = X_2 , \quad v = - X_3\cos\omega + X_1\sin\omega , \quad W = X_1\cos\omega + X_3\sin\omega ,$$

where $\omega = \sin^{-1}(r\sin\varphi/R)$. The distribution of the components u, v and w over the hole for a tensile load of 1.18 kN is depicted in Fig.3.6. The error in the absolute fringe-order determination is assumed to be 0.15 fringe and equal for all the viewing directions. In this case, the absolute errors on the displacement components do not exceed 0.2 μm for w, and 0.3 μm for u and v.

For reflection holograms, the best interferometer design has the parameters $\alpha_1 = 0$, $\alpha_2 = \alpha_3 = 60°$, and $\psi_1 = \psi_2 = \psi_3 = \psi$ (Fig.3.3). It is based on the quasi-orthogonal matrix

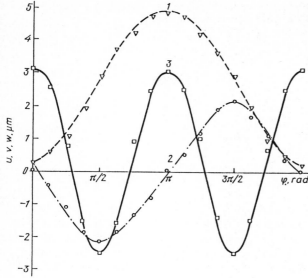

Fig.3.6. Distribution of displacement components along the edge of a hole in a shell under tension: (*1*) u, (*2*) v, and (*3*) w

$$T_3 = A_0 \begin{bmatrix} 1/\sqrt{3} & -\sqrt{2}/\sqrt{3} & 0 \\ 1/\sqrt{3} & \sqrt{2}/2\sqrt{3} & -\sqrt{2}/2 \\ 1/\sqrt{3} & \sqrt{2}/2\sqrt{3} & \sqrt{2}/2 \end{bmatrix} \qquad (3.47)$$

and a scaling matrix D_2 with the elements

$$A = \sqrt{3}(1 + \cos\psi) , \quad B = (\sqrt{3}/2)\sin\psi .$$

In this case the viewing directions lie on the generatrices of a cone with the apex at the origin and the illumination direction $\hat{e}_S$ coincides with the cone's axis.

This interferometer setup was implemented in the following way [3. 28]. A two-exposure hologram recorded in an opposed-beam arrangement is fixed in the x_2,x_3 plane (Fig.3.7) and illuminated with a collimated white-light beam along the x_1 axis. A telecentric viewing system is mounted in the x_1,x_2 plane at an angle ψ to the x_1 axis. The fringe patterns are recorded on a photographic plate in the initial position and two positions corresponding to two consecutive rotations of the plate in the x_2,x_3 plane through 120° about the x_1 axis.

An essential advantage of this interferometer configuration over the multi-hologram setups is the possibility of a visual determination of the angle ψ in the course of the plate's rotation which permits observation of the part of the object surface under study from all the three viewing directions. Increasing the angle ψ dramatically reduces the error on the "in-plane" components X_2 and X_3 while the "out-of-plane" component X_1 error remains nearly constant.

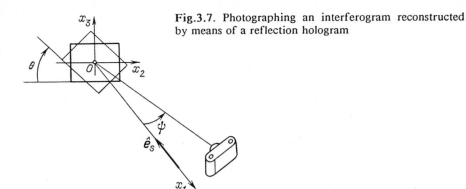

Fig.3.7. Photographing an interferogram reconstructed by means of a reflection hologram

Another remarkable feature of this setup is that when all the viewing directions are simultaneously rotated through any angle around the x_1 axis, we obtain a C-optimal system [3.28]. Thus, additional information can be derived and the measurement accuracy improved. This is done by recording the fringe patterns corresponding to the photographic plate rotated around the x_1 axis in the x_2, x_3 plane through an angle $\theta = (2\pi/3m)$ radians (m is an integer). These patterns allow obtaining m systems of equations, each corresponding to an optimum interferometer arrangement. Assuming that the errors in the fringe order Δn_j (j = 1, 2, 3) are known and equal to Δn which is a random quantity with a Gaussian distribution and a zero mean, the error estimate derived from (3.8) will coincide with the upper bound on the error obtained with (3.34).

An optical system using reflection holograms was employed to study the deformation of the edge of a circular hole in a cylindrical shell (Fig.3.4) subjected to a torsional stress. The stability of loading of the shell is shown by the multi-exposure interferogram shown in Fig.3.8. Figure 3.9a, b and c presents three-exposure interferograms of the cylindrical shell (Sect.2.3.3) loaded by a torque M = 17.3 N·m for the viewing directions $\hat{e}_1$, $\hat{e}_2$ and $\hat{e}_3$, respectively. Image reconstruction was carried out in white light for $\psi = 37.5°$. The brightest fringes in the interferograms are the zero-motion

Fig.3.8. Multi-exposure interferogram of a cylindrical shell with a circular hole under torsion

106

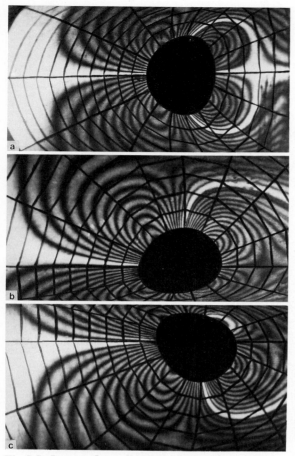

Fig.3.9. Image of a cylindrical shell under torsion reconstructed with one reflection hologram in the (a) $\hat{e}_1$, and (b) $\hat{e}_2$, and (c) $\hat{e}_3$ directions

fringes. The interferograms were recorded with the hologram rotated through $\theta = 30°$. This permitted construction of four systems of equations of the type (3.29) satisfying the C-optimality criterion for each point on the edge of the hole. Figure 3.10 gives average displacement components u, v and w along the edge of the hole. The error in the determination of the w component was found to be 0.07 μm, and 0.1 μm for u and v.

3.2.3 Determination of Two Displacement-Vector Components

The C-optimality criterion can be used to design an interferometer to measure two components of the displacement vector, one of them normal, and the other tangential to the object surface [3.29].

For known absolute fringe orders, the interferometer arrangement then contains one illumination direction $\hat{e}_S$ along the x_1 axis and two viewing

107

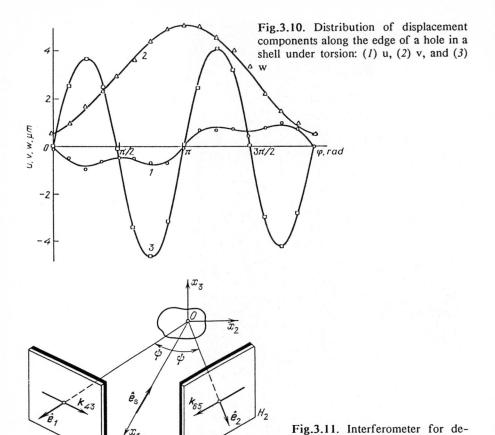

Fig.3.10. Distribution of displacement components along the edge of a hole in a shell under torsion: (*1*) u, (*2*) v, and (*3*) w

Fig.3.11. Interferometer for determination of two displacement vector components

directions, $\hat{e}_1$ and $\hat{e}_2$, at an angle ψ to the x_1 axis in the x_1, x_2 plane (Fig.3.11). The 2×2 sensitivity matrix K_N of such an interferometer can be written as

$$K_N = \begin{bmatrix} 1+\cos\psi & -\sin\psi \\ 1+\cos\psi & \sin\psi \end{bmatrix}. \qquad (3.48)$$

One can use a scaling matrix D_2

$$D_2 = \begin{bmatrix} (1+\cos\psi)^{-1} & 0 \\ 0 & (\sin\psi)^{-1} \end{bmatrix}$$

to transform the sensitivity matrix (3.48) to the orthogonal form

$$K_N^* = \begin{bmatrix} 1 & -1 \\ 1 & 1 \end{bmatrix}.$$

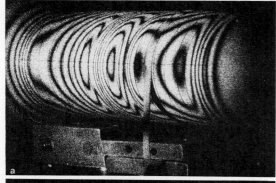

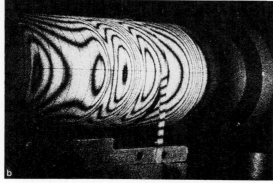

Fig.3.12. Reconstructed image of a cylindrical shell with an inner circular groove obtained by viewing in (a) $\hat{e}_1$ and (b) $\hat{e}_2$ directions given in Fig.3.11

By following a reasoning similar to that in Sect.3.2.2, one can estimate the accuracy in determinating two displacement vector components if a point source of light is used and the observation points are located at a finite distance from the object.

Figure 3.12a,b presents interferograms of a cylindrical shell with an inner circular groove loaded by an inner pressure of 1 MPa and a tensile stress of 2.9 kN. The shell, made of D16T alloy, has an outer diameter of 60 mm with a wall 2 mm thick. The circular groove, 1 mm deep and 20 mm wide, was at the center of the shell section under study. The distribution of the displacement components $X_2 = U$ and $X_1 = W$ along the cylinder generatrix shown in Fig.3.13 gives evidence for a strong effect of the groove on the radial displacements.

When using the relative fringe orders to interpret the interferograms, the optimal sensitivity matrix $K_{\delta N}$ takes on the form:

$$K_{\delta N} = \frac{2\ell}{\sqrt{\ell^2 + R^2}} \begin{bmatrix} \sin\psi & \cos\psi \\ \sin\psi & -\cos\psi \end{bmatrix} \qquad (3.49)$$

where ℓ is the distance between the observation points, and R is the distance from hologram center to the origin. The transitions between the observation points on the holograms H_1 and H_2 corresponding to this matrix and defined by the sensitivity vectors k_{43} and k_{65} are shown in Fig.3.11.

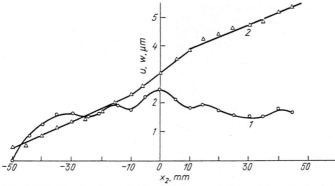

Fig.3.13. Distribution of (*1*) radial w, and (*2*) axial u displacement components along the cylinder generatrix

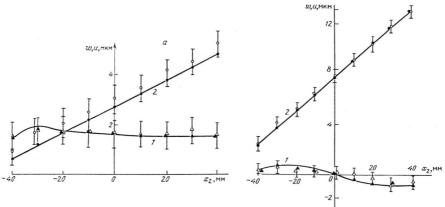

Fig.3.14. Distribution of (*1*) radial and (*2*) tangential displacement components along the generatrix with the shell subjected to (**a**) uniform pressure and tension, and (**b**) tensile stress only: ●▲ from absolute order number, ○△ from relative order numbers

Since in a two-hologram interferometer one can implement, in a single experiment, optimum configurations for the cases where the holograms can be interpreted through either absolute or relative fringe orders, it allows an experimental comparison of these two methods. This comparison has been made by *Pryputniewicz* and *Bowley* [3.30] for the case of an object translation, but the selection of the interferometer parameters was not validated in any way.

Figure 3.14 illustrates determination of the in-plane and radial components of the displacement vector along the generatrix of a smooth cylindrical shell loaded by a tensile stress and an inner pressure [3.29]. By properly varying the loads, one can change the relative magnitudes of the in-plane and radial displacements. The curves in Fig.3.14a correspond to a simultaneous loading of the shell by an inner pressure of 0.30 MPa and a tensile stress of 1.7 kN, and in Fig.3.14b only by a tensile stress of 2.9 kN. The relative fringe orders were determined by scanning the holograms with a

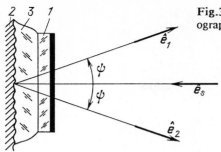

Fig.3.15. Schematic diagram of an overlay holographic interferometer

nonexpanded laser beam. For a small in-plane displacement component (Fig.3.14a) the maximum relative fringe order was three, and for a large one (Fig.3.14b) it was ten. As seen from the graphs, the displacement components derived by the two methods of hologram interpretation coincide within an experimental error.

Despite the fact that the relative fringe-order method provides equal accuracy for the displacement components, it is too labor-intensive to be used in practical work and thus can only be recommended for measurements in cases where the number of points on the object surface studied is not large.

3.2.4 Displacement Measurements with an Overlay Interferometer

In strain studies of massive elements with reflection-hologram interferometers the recording medium is fixed to the object under investigation with mechanical or magnetic devices [3.31]. These devices cannot be employed with thin-walled objects, however, since they can distort the strain field of interest.

In the case of plane surface objects one can fix the photographic plate in place by means of an intermediate transparent optical medium [3.32]. This medium should possess a low piezo-optical sensitivity and a low shear rigidity compared to the substrate of the plate, and maintain constant opto-mechanical properties in the course of the experiment. The photographic plate should not displace relative to the object surface. When the recording medium is fixed in such a way, we have the so-called overlay interferometer.

The overlay interferometer configuration is shown in Fig.3.15. Photographic plate *1* is attached to the surface of the object *2* through an intermediate medium *3*, usually synthetic rubber. Synthetic rubber greatly reduces the weight of the plate fixture and has almost no effect on the object deformation due to its low rigidity.

Quantitative determination of displacements can be conveniently carried out with an interferometer arrangement where the object is illuminated with a plane wave at normal incidence and the observation is carried out from four directions aligned symmetrically with respect to the surface normal (Sect.2.3.2).

111

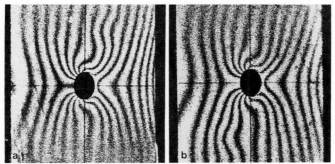

Fig.3.16. Fringe pattern obtained with an overlay interferometer in (a) $\hat{e}_1$ and (b) $\hat{e}_2$ directions (courtesy of S.I. Gerasimov)

The possibility of obtaining high-quality interferograms by means of an overlay interferometer is illustrated by Fig.3.16a,b showing fringe patterns for a plane specimen with a central hole subjected to a tensile stress. These interferograms were obtained from two viewing directions symmetrically spaced about the surface normal.

4. Determination
of the Displacement-Vector Components

The methods for the interpretation of holographic interferograms discussed in the preceding chapter permit determination of all three components of the displacement vector X_1, X_2 and X_3 of an arbitrary point on the surface of a strained object. The displacement vector components normal and tangential to the object surface, which are necessary for the analysis of the strained state can, in general, be derived by calculation, too. However, for plane objects, the holographic interferometry is capable of recording individual components of the displacement vector as well. In this chapter we discuss the various methods for separate determination of the in-plane and out-of-plane displacements.

4.1 Determination of Out-of-Plane Displacements

4.1.1 Interferometer Systems

As follows from the principal relation of holographic interferometry (2.47), the sensitivity vector coincides with the bisector of the angle made by the illumination and viewing directions. The sensitivity vector can be brought into coincidence with the normal to the object surface by two techniques. In the first, the illumination and viewing directions are at equal angles to the surface normal. In the second, the illumination and viewing directions coincide with the surface normal so that the interferometer is insensitive to the in-plane component of the displacement vector and, at the same time, possesses maximum sensitivity for the out-of-plane component. An optical system satisfying this requirement is shown in Fig.4.1a [4.1, 2].

Laser beam *1* expanded by collimator *2* impinges on the beam splitter *3*. The plane wave reflected from the splitter illuminates the surface of the object *4*. This wave reflected from the object is incident on the photographic plate *5* (object wave). The wave passing through the beam splitter is reflected by the mirror *6* onto the photographic plate *5* (reference wave). The Cartesian coordinate system (x_1, x_2, x_3) is chosen such that the x_1 axis is directed along the normal to the object surface, and the x_2 and x_3 axes lie in its plane. Under these conditions the direction of illumination defined by the unit vector $\hat{e}_S$ will be parallel to the x_1 axis and, hence, normal to the object surface.

The reconstructed image is viewed by means of the optical system exhibited in Fig.4.1b. When the hologram *1* is illuminated by the reference wave, the image of the object is viewed on the screen *2* through a telecen-

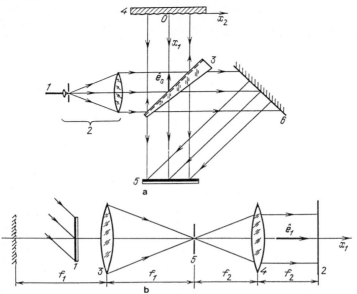

Fig.4.1. Interferometer setup for determination of out-of-plane displacements: (a) hologram recording, (b) observation of the image

tric system made up of the lenses *3* and *4*. The diaphragm *5* placed in the lens focus ensures spatial filtering of the object wave, as a result of which the viewing direction defined by the unit vector $\hat{e}_1$ will also be parallel to the x_1 axis everywhere on the object surface.

With such an optical arrangement, the interference fringes observed on the object image and obtained, for instance, by the two-exposure method can be interpreted mechanically as loci of equal out-of-plane components of the displacement vectors.

Substituting the components of the vectors

$$\hat{e}_1 = (1,0,0) , \quad \hat{e}_S = (-1,0,0) , \quad d = (X_1,X_2,X_3)$$

into (2.44) yields an expression for the out-of-plane components

$$X_1 = n\lambda/2 , \tag{4.1}$$

where n is the absolute order of the bright fringe at the point in question, and λ is the wavelength of light.

Such an interferometer system has the remarkable capability to visualize out-of-plane displacement components in the presence of very large in-plane components sometimes exceeding the former by an order of magnitude [4.2]. It should be kept in mind, however, that the fringe contrast becomes degraded as the in-plane displacement components grow. The area of the object surface which can be studied with such systems is limited by the size of the collimating lens, typically not exceeding 200 mm.

114

De Larminot and *Wei* [4.1] proposed an optical arrangement for the visualization of out-of-plane displacements in the presence of considerable rigid translations and tilting of the object under study. When observing a fringe pattern in real time, the translations of the object are compensated by displacing the hologram, and its rotation, by tilting the beam splitter. This approach was employed for the visualization of contraction of specimens with cracks and circular holes subjected to a tensile stress.

Such a system is also used for the visualization of the out-of-plane displacement components in a pure specimen bending. This permits determination of Poisson's ratio for the given material. In pure bending of a plate, the surface acquires an anti-clastic shape (Sect.5.2.1), with the constant deflection lines forming a family of hyperbolas described by [4.3]

$$x_2^2 - \mu x_3^2 = \text{const.} \quad .\tag{4.2}$$

For the equation of the asymptotes to these hyperbolas we have

$$x_2^2 - \mu x_3^2 = 0 \; .\tag{4.3}$$

From (4.3) one can derive Poisson's ratio

$$\mu = \tan^2 \alpha \; ,\tag{4.4}$$

where α is one half of the angle between the hyperbola's asymptotes. By this method, classical interferometry permits the determination of Poisson's ratio only on specimens with specular surfaces [4.4], whereas holographic interferometry also allows studies of specimens with diffusely reflecting surfaces, thus substantially broadening the available range of materials which can be studied [4.5,6]. A typical two-exposure interferogram obtained on a $2 \times 30 \times 100$ mm^3 specimen made of D16T alloy and subjected to a four-point bending is presented in Fig.4.2. For this material, Poisson's ratio $\mu = 0.31 \pm 0.05$.

In cases where one knows a priori that the displacement vector is normal to the object surface (as is usual, for instance, in plate bending) the displacement can be determined by means of an optical system with the spherical illumination wave as depicted in Fig.4.3 [4.7]. The plate is illuminated with the wave from a point source S while observation is done from a point B at a finite distance away from the surface. To determine the displacements X_1 at various points on the object surface (x_2, x_3) with due account of the variation of the sensitivity vector, one can use the expression

$$X_1 = \frac{n\lambda}{[1+(x_2^2+x_3^2)/x_{1S}^2]^{-1/2} + [1+(x_2^2+x_3^2)/x_{1B}^2]^{-1/2}} \; ,\tag{4.5}$$

where n is the absolute fringe order at the surface point in question; $(x_{1S},0,0)$ and $(x_{1B},0,0)$ are the coordinates of the point source S and observation point B, respectively. Note that for x_2, $x_3 \ll x_{1S}$, and x_2, $x_3 \ll x_{1B}$ (4.5) transforms into (4.1), in which case the variation of the sensitivity

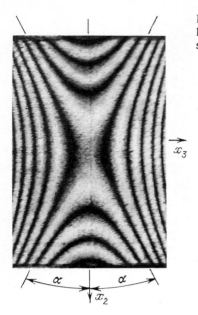

Fig.4.2. Reconstructed image on a two-exposure hologram of a plane specimen under pure bending stress

vector over the object surface may be disregarded. This optical system permits to study deformation of objects that are larger than those that can be examined with the system shown in Fig.4.1.

The interferometer in Fig.4.3 can be used in strain-compensation measurements permitting determination of Poisson's ratio of a specimen in pure bending [4.8]. The measurements are made on real-time interferograms in the following order:
• a plane-parallel plate is placed in front of the beam splitter as in the arrangement in Fig.4.3a;
• a hologram of the initial state of the specimen is obtained and processed after which it is placed back into the exact position it occupied for the initial exposure;
• the plane-parallel plate is mounted behind the beam splitter in the path of the reference wave, resulting in the formation of concentric interference fringes on the specimen's surface of a system corresponding to a fictitious axially symmetrical strain;
• the bending moment M_1 is found at which fringes parallel to the specimen's axis are observed (Fig.4.3b), thus showing compensation of the stretch;
• the bending moment M_2 is determined under the conditions where interference fringes orthogonal to the specimen's axis are observed (Fig.4.3c), which implies compensation of the lateral strain.
The ratio of the bending moments M_1 and M_2 yields Poisson's ratio for the specimen.

This technique not only broadens the range of Poisson's ratios which can be measured, it also has greater accuracy than the previously described method of Fig.4.2. Indeed, by this method the measurement error for Poisson's ratios on the order of 0.1 does not exceed 5% [4.8].

116

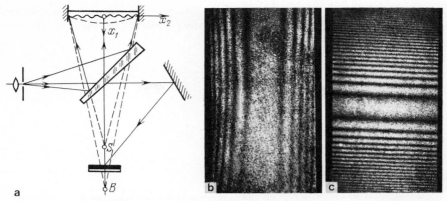

Fig.4.3. Determination of out-of-plane displacements with a spherical illuminating wave: **(a)** optical system; **(b)** compensation for longitudinal strain; **(c)** compensation for lateral strain

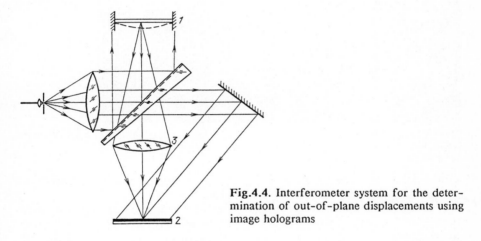

Fig.4.4. Interferometer system for the determination of out-of-plane displacements using image holograms

Figure 4.4 presents an interferometer system for the visualization of out-of-plane displacements which uses focussed image holograms [4.6]. The surface of object *1* is imaged onto the plane of hologram *2* by means of objective lens *3*. As already pointed out, in this case the fringes localize in the hologram plane thus permitting broadening of the range of measurable displacements.

These interferometer systems have been used widely in studies of the deformation of plates and cantilever beams [4.9-11]. Holographic interferometry is particularly useful to study the bending of plates representing multiply connected regions whose theoretical analysis meets with formidable difficulties. Figure 4.5 shows a reconstructed image of a circular ring of constant thickness which is fixed at the outer and inner edges, the radius of the inner circle being one half that of the outer one. The ring is loaded by a uniform pressure [4.12]. A two-exposure interferogram was obtained

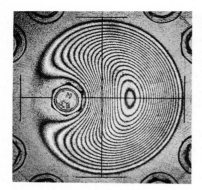

Fig.4.5. Reconstructed image of a plate whose working area is a doubly connected region

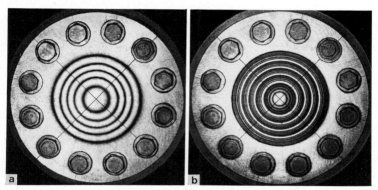

Fig.4.6. (a) Two-exposure and (b) multi-exposure interferogram of a loaded plate held fixed at its edge

with the optical system shown in Fig.4.3. The fringe pattern is essentially a topogram of the strained surface made with a step of $\lambda/2$.

Large deflections can be measured by the technique proposed by *Hsu* [4.13] involving illumination of the object by a mirror which can be rotated about an axis lying in its plane. If one observes the interference pattern of the strained object in real time, then the number of the fringes on the object can be substantially reduced by adjusting the angle of the mirror. By knowing the angle of the mirror and the fringe order in the reconstructed image of the object, one can readily derive the displacements. Since this represents essentially a compensation technique, it is very efficient for the measurement of large gradient deflections.

In the case of small deflections (on the order of λ) one can improve the sensitivity by using a technique which is based on superposition of conjugate diffraction orders produced by illuminating a nonlinearly recorded hologram [4.14].

The accuracy of plate-deflection measurement can be improved by employing multi-exposure holographic interferometry (Sect.2.1.4). Figure 4.6a displays a two-exposure holographic interferogram of a circular plate fixed at the edge and loaded by a uniformly distributed load, and Fig.4.6b a multi-exposure hologram recorded with four consecutive load increments. In both cases the loads are increased by the same amount. The narrowing of the bright fringes facilitates determination of the coordinates for their centers. Besides this observation, the presence of narrow bright fringes in the reconstructed image demonstrates the stability of the plate to loading.

4.1.2 Determination of the Sign of Displacement

Identification of the sign (i.e., direction) of a displacement normal to an object surface is ambiguous. This is because the change in the phase associated with the deformation of an object is an argument of the characteristic fringe function squared, see (2.32). In some specific cases additional information is available, which describes the actual character of deformation, for example, the direction of loading. In the general case, however, one cannot derive the sign of displacement from the fringe pattern. This situation can arise in determining residual stresses (Sect.6.2) where the direction of the out-of-plane displacement component is not unambiguously connected with the sign of the stresses being measured.

Two major methods for determining the sign of displacement have currently found application in double-exposure holographic interferometry. One of them is based on shifting the strained object in a known direction between the exposures [4.15]. The additional shift is caused by turning either the object or the hologram, whichever is more convenient, about an axis lying in the plane of its surface. Focussed image holograms should also be rotated about an axis lying in their plane. In the Leith–Upatnieks configuration, the hologram is turned about an axis lying at a finite distance from it (Sect.4.2.1). An increase or decrease in the number of fringes compared with the original unshifted pattern defines the direction of the displacement. Since deformation-induced out-of-plane displacements produce fringes localized on the object surface, the additional shift should likewise cause the appearance of fringes on the surface.

Consider the application of this method to determine the sign of deflection of a cantilever beam. Figure 4.7a depicts an initial double exposure interferogram of the deflection of a cantilever beam. Figure 4.7b displays a reconstructed image of the beam observed after applying the same load between the exposures but for the second interferogram, the beam was turned about an axis on its surface and perpendicular to the beam axis. The direction of rotation is known and its amount is checked by the fringes observed on the witness, in this case a nondeformed plate located to the left of the beam and fixed rigidly to its base. Since the number of the fringes on the beam has decreased after its rotation, the sign of the deflection should be opposite to that of its rotation. In the above example, two interferograms with one increment of loading were recorded by means of a kinematic device (Fig.2.4) designed to precisely replace holograms back to the original position of exposure. For this purpose different parts of the same holograms, or two reference waves or special optical systems with image splitters (described in Chap.6) can be used, too.

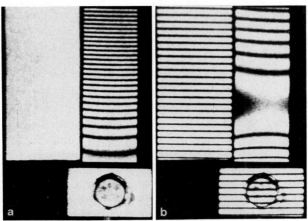

Fig.4.7. Reconstructed images of a deflected cantilever beam in (a) infinite, and (b) finite width fringes

Another method for determining the sign of an out-of-plane displacement is based on the variation of the fringe pattern that occurs when the parameters of the optical system are changed at the reconstruction step of the image which is recorded on a doubly exposed hologram. Here, one can conveniently employ the optical system of fringe control displayed in Fig.2.33b provided it has been preliminarily calibrated in the direction of displacement of the illumination source. This method requires recording of only one hologram.

In the real-time method, additional displacements of a known sign are made after the object is loaded. The change of the original fringe pattern observed under these conditions permits determination of the sign of the displacement exactly as in the two-exposure technique. The sign of displacement in the real-time method can be determined either from an analysis of the fringe dynamics under a smooth variation of the load (Sect. 2.3.3), or using the fringe-control approach (Fig.2.23a).

The simplest way to determine the sign of a displacement is by employing sandwich interferometry [4.16]. The change in the frequency of the fringes on the object surface observed by turning the sandwich hologram gives the sign of the displacement.

4.2 Measurement of In-Plane Displacements

4.2.1 Moiré Contouring Techniques

When the displacement vector is known a priori to lie in the plane of the object surface, it must be measured with an interferometer setup in which the sensitivity vector has a component parallel to the surface [4.17]. To measure in-plane displacements in the presence of out-of-plane movement, it is required that the sensitivity vector be strictly tangential to the surface

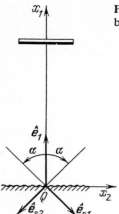

x_1

$\hat{e}_1$

α α

O

x_2

$\hat{e}_{S2}$ $\hat{e}_{S1}$

Fig.4.8. Interferometer setup with two collimated illumination beams for the observation of moiré fringes

of the object under investigation. However, since this vector is defined as the difference between the unit vectors of observation and illumination (2.46), designing such interferometers is impossible. Therefore, the measurement of in-plane displacement components by holographic interferometry represents a more involved problem than that of measuring the out-of-plane components.

One of the first interferometer setups intended for the determination of in-plane displacement components at individual points was proposed by *Ennos* [4.18]; it is described in Sect.2.3.2. This system uses two viewing directions spaced symmetrically about the surface normal. Based on the fact that illumination and viewing points are interchangeable, *Boone* [4.19] proposed an interferometer setup with two directions of illumination and one viewing direction. Each direction of illumination produces its own fringe pattern which, when observed simultaneously, create moiré fringes which contain information only on the in-plane displacement-vector component. This method of the *moiré contouring* was subsequently improved [4.20-22].

Consider the formation of holographic moiré fringes [4.21]. Let an object with a plane surface be illuminated with two collimated light beams whose directions are specified by the unit vectors $\hat{e}_{S1}$ and $\hat{e}_{S2}$ lying symmetrically about the normal to the object surface (Fig.4.8). A telecentric system views the surface in the direction of the vector $\hat{e}_1$ which is normal to it. The Cartesian coordinate system (x_1, x_2, x_3) is chosen such that the x_1 axis coincides with the viewing direction, and the x_2 and x_3 axes lie in the surface plane. We also assume that the vectors $\hat{e}_{S1}$, $\hat{e}_{S2}$ and $\hat{e}_1$ lie in the x_1, x_3 plane. Holographic fringe patterns of a strained object that is illuminated simultaneously from two directions, are obtained by the two-exposure technique.

The principal relation of holographic interferometry (2.44) can be written for each direction of illumination, i.e.,

$$(\hat{e}_1 - \hat{e}_{S1}) \cdot \mathbf{d} = n_1 \lambda , \tag{4.6}$$

$$(\hat{e}_1 - \hat{e}_{S2}) \cdot \mathbf{d} = n_2 \lambda , \tag{4.7}$$

121

where **d** is the displacement vector of the point of interest, and n_1 and n_2 are the absolute fringe orders at this point. When reconstructing the light waves recorded on a hologram, the interference fringes described by (4.6, 7) are observed simultaneously. Subtracting (4.7) from (4.6) yields a parametric equation for the moiré fringes thus formed:

$$(\hat{e}_{S2} - \hat{e}_{S1}) \cdot d = -n_m \lambda , \qquad (4.8)$$

where $n_m = (n_2 - n_1)$ is the *moiré-fringe order*. The components of the $\hat{e}_{S1}$, $\hat{e}_{S2}$ and **d** vectors in our coordinate system have the form

$$\hat{e}_{S1} = (-\cos\alpha, \sin\alpha, 0) , \quad \hat{e}_{S2} = (-\cos\alpha, -\sin\alpha, 0) ,$$

$$d = (X_1, X_2, X_3) , \qquad (4.9)$$

where α is the angle between the illumination and viewing directions. Substituting (4.9) into (4.8) gives the expression for deriving the in-plane displacement component X_2 from the moiré pattern:

$$X_2 = n_m \lambda / 2\sin\alpha . \qquad (4.10)$$

The moiré fringes characterizing the other component X_3, are obtained in the same way by placing the illuminating beams in the x_1, x_3 plane.

Because of the low density of the fringe patterns caused by an object's deformation, the moiré fringes usually have low contrast. The contrast can be improved by fringe multiplication which can usually be achieved by adjusting the various parameters of the interferometer system between the exposures. The interference fringes formed in this way are called auxiliary and should satisfy the following requirements:
• the fringe spacing should be the same for the two illuminating beams;
• the interference fringes should be placed on the object surface.
The auxiliary fringe pattern can be produced by rotating the object [4.23], by changing the direction of illumination [4.24], or by rotating the hologram [4.24-26]. The latter technique turns out to be the most convenient.

When illuminating a hologram recorded in the Leith–Upatnieks configuration, the interference fringes which appear when the photographic plate is rotated about an axis parallel to the plate's plane, localize on the object surface, provided that [4.21]

$$x_{10} = L(1 + \cos\theta_r) , \qquad (4.11)$$

where x_{10} is the distance from the axis of rotation to the hologram plane, L is the hologram-to-object distance, and θ_r is the angle between the reference beam and the normal to the hologram surface. Fringe multiplication with focussed image holograms is also done by rotating the photographic plate about an axis lying in its plane.

The application of the moiré contouring method can best be illustrated by considering the determination of in-plane displacements on the surface of a disc loaded by concentrated forces. The optical system produces focus-

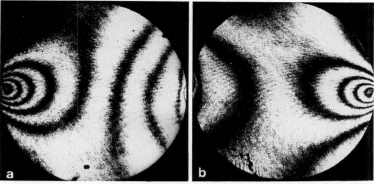

Fig.4.9. Fringe patterns observed with a loaded disc illuminated from (a) $\hat{e}_{s2}$ and (b) $\hat{e}_{s1}$ directions

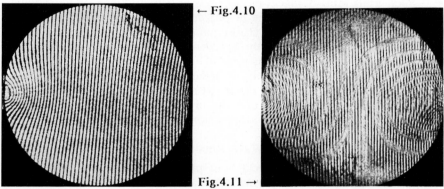

← Fig.4.10

Fig.4.11 →

Fig.4.10. Interferogram of a deformed disc where the photographic plate has been rotated for the illumination, producing fringe multiplication

Fig.4.11. Moiré pattern characterizing in-plane displacement distribution for the disc

sed image holograms. The directions of illumination, the disc's surface normal, and the directions of the forces all lie in the same plane. With such an orientation of the illuminating beams, the in-plane displacement vector components can be measured in the direction of the loading. Figure 4.9a,b presents the fringe patterns corresponding to one loading step and obtained separately for each illumination direction. The small number of the fringes available does not permit obtaining clearly visible moiré fringes under simultaneous observation. A fringe pattern caused by the disc's deformation and the photographic plate's rotation about an axis lying in its plane and in the direction orthogonal to the action of the forces is depicted in Fig.4.10; it graphically demonstrates the effect of fringe multiplication. Figure 4.11 presents a moiré pattern characteristic of this case and obtained by viewing a real image in white light. The observed moiré pattern is in agreement with the theoretically calculated lines of equal in-plane displacement. The fringe contrast can be improved by using spatial filtering [4.21].

An attempt has been made to apply holographic moiré contouring to measuring in-plane displacements of strained cylindrical shells [4.26]. By using holographic moiré contouring in combination with the method of out-of-plane displacement visualization, discussed in Sect.2.1.1, one can measure two displacements in one experiment [4.25].

Another approach in the holographic moiré technique makes use of a standing interference field [4.27, 28]. In experimental mechanics one frequently employs sinusoidal high-frequency gratings which can be easily produced by recording an interference pattern in the region of intersection of two plane waves. Fabrication of an object raster usually requires coating the object surface under study with a high-resolution emulsion, photoresist, etc. A standing interference field can also be employed as a reference raster. In this case the moiré fringes are observed in real time.

Such a standing fringe field is used also to produce high-frequency metallized rasters on the surface of objects under study ($\simeq 1000$ lines/mm). The recording of holographic fringe patterns with such rasters in Denisyuk's configuration substantially broadens the possibilities of the moiré technique [4.29, 30]. Reconstruction of the waves recorded on such a doubly-exposed hologram permits an accurate comparison of the replicas of the original and the deformed rasters. It allows to determine the angles of rotation of the object surface and the in-plane displacements by viewing the hologram in reflection and transmission, respectively.

4.2.2 Method of Reference-Wave Reconstruction

In-plane displacement of objects with plane surfaces can also be determined with a method based on the reference-wave reconstruction [4.31, 32]. In the first step a focussed image hologram of an object illuminated by two plane waves oriented symmetrically about the normal to its surface is recorded. After processing and returning the hologram back to the site of its exposure, it is illuminated only with object waves, thus reconstructing the reference. After deformation, the intensity distribution in the reference wave is modulated by fringes representing loci of equal in-plane displacements.

We shall derive the principal relation following the approach of *Katziz* and *Glaser* [4.32]. In the optical configuration for recording focussed image holograms (Fig.4.12a), the unit illumination vectors $\hat{e}_{S1}$ and $\hat{e}_{S2}$ and the viewing vector $\hat{e}_1$ lie in one plane. The object and the photographic plate are illuminated by plane waves. We denote the complex amplitudes in the hologram plane by A_1 and A_2, and that of the reference by A_r. Then, for the light intensity distribution I in the hologram plane we obtain

$$I = \left| A_1 + A_2 + A_r \right|^2$$

$$= \left| A_r \right|^2 + \left| A_1 + A_2 \right|^2 + (A_1 + A_2)A_r^* + (A_1 + A_2)^* A_r , \qquad (4.12)$$

where the asterisk refers to a complex conjugate.

124

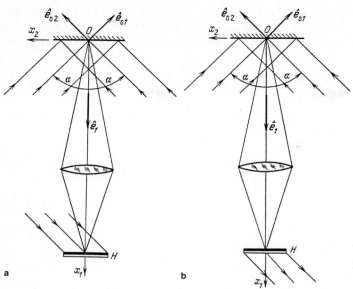

Fig.4.12. Generation of focussed image hologram (a) with the object illuminated by two beams, and (b) in the reconstruction of the reference wave

We assume that the hologram H, which has an amplitude transmission after photographic processing and is proportional to the intensity in (4.12), is returned precisely to the same exposure position. If we now illuminate the hologram with object waves only (Fig.4.12b), then we obtain for the expression of the complex amplitude A behind the hologram

$$A = I(A_1 + A_2) = (|A_r|^2 + |A_1 + A_2|^2)(A_1 + A_2)$$
$$+ (A_1 + A_2)^2 A_r^* + |A_1 + A_2|^2 A_r . \tag{4.13}$$

Since we are interested in the light wave propagating in the direction of the reference, we consider only the third term on the right-hand side of (4.13), which, for convenience, can be denoted by A_0

$$A_0 = |A_1 + A_2|^2 A_r$$
$$= (|A_1|^2 + |A_2|^2)A_r + (A_1 A_2^* + A_1^* A_2)A_r . \tag{4.14}$$

Since the reference wave is planar, and the waves A_1 and A_2 are not correlated, the second term in (4.14) may be neglected:

$$A_0 = (|A_1|^2 + |A_2|^2)A_r = (A_1^* A_1 + A_2^* A_2)A_r . \tag{4.15}$$

This expression describes the reference wave used in recording to within a multiplicative constant.

We now let the object deform in such a way that the displacements of points on its surface do not exceed the minimum speckle size determined

125

by the lens aperture and the magnification. Then the amplitudes of the new object waves A_1' and A_2' incident on the hologram will differ from the original ones only in phase. Substituting A_1' and A_2' for A_1 and A_2 in (4.15) we obtain

$$A_0' = (A_1{}^*A_1' + A_2{}^*A_2')A_r .$$ \hfill (4.16)

The complex object wave ampltidues A_1, A_2, A_1' and A_2' can be written in the following form

$$A_1 = a_1 , \quad A_2 = a_2 ,$$
$$A_1' = a_1 \exp[-i(2\pi/\lambda)(\hat{e}_1 - \hat{e}_{S1}) \cdot d] ,$$
$$A_2' = a_2 \exp[-i(2\pi/\lambda)(\hat{e}_1 - \hat{e}_{S2}) \cdot d] .$$ \hfill (4.17)

For the coordinate system (x_1, x_2, x_3) shown in Fig.4.12b, the components of the vectors $\hat{e}_{S1}$, $\hat{e}_{S2}$, $\hat{e}_1$ and d will be

$$\hat{e}_{S1} = (-\cos\alpha, -\sin\alpha, 0) , \quad \hat{e}_1 = (1, 0, 0) ,$$
$$\hat{e}_{S2} = (-\cos\alpha, \sin\alpha, 0) , \quad d = (X_1, X_2, X_3) .$$ \hfill (4.18)

Substituting (4.18) into (4.17) yields

$$A_1' = a_1 \exp\{-i(2\pi/\lambda)[X_1(1+\cos\alpha) + X_2\sin\alpha]\} ,$$
$$A_2' = a_2 \exp\{-i(2\pi/\lambda)[X_1(1+\cos\alpha) - X_2\sin\alpha]\} .$$ \hfill (4.19)

By using (4.19) one can determine the square of the modulus of (4.16)

$$|A_0'|^2 = 4a_0{}^4 \cos^2[(4\pi/\lambda)X_2\sin\alpha] .$$ \hfill (4.20)

Thus, the intensity of the reconstructed reference wave is modulated by cosinusoidal fringes of equal in-plane displacements X_2 . Their magnitude can be found from

$$X_2 = n\lambda/4\sin\alpha ,$$ \hfill (4.21)

where n is the fringe order. The contrast of the fringes described by (4.20) is unity.

A remarkable feature of focussed-image holograms is the existence of a one-to-one correspondence between the points of the image and of the object surface. Hence, for each point in the reconstructed reference wave one can find the corresponding point on the object's image. In other words, in the field of a reconstructed reference wave one can identify points on the object's surface. By comparing (4.21 and 10) one sees that the reference-wave reconstruction method is twice as sensitive as the holographic moiré technique.

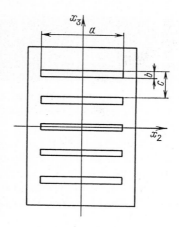

Fig.4.13. Interferogram of a diametrically compressed disc obtained by reference wave reconstruction

Fig.4.14. Slit mask for in-plane displacement vector measurements

Figure 4.13, obtained by the above method, presents an interferogram of the surface of a disc loaded by concentrated forces. The fringe pattern is seen to be similar to that produced by moiré contouring (Fig.4.11). The reference-wave reconstruction method can be used in combination with the technique discussed in Sect.4.1.4 (Fig.4.4) to simultaneously measure the in-plane and out-of-plane displacements [4.31].

4.2.3 Determination of In-Plane Displacements from the Fringe Visibility

The fringe visibility in the holographic interferometry of diffusely reflecting objects depends on the aperture of the observation system employed, and the out-of-plane components of the displacement vector. Let us place a slit mask in front of a doubly exposed hologram used to reconstruct a real image (Fig.4.14). One can readily show that the fringe visibility V is described by [4.33]

$$V = \left| \mathrm{sinc}\left(\frac{\pi a X_2}{\lambda x_1}\right) \mathrm{sinc}\left(\frac{\pi b X_3}{\lambda x_1}\right) \frac{\sin(m\pi c X_3/\lambda x_1)}{m\sin(\pi c X_3/\lambda x_1)} \right|, \tag{4.22}$$

where a, b are the length and width of one slit, c is the slit spacing, m is the number of slits, x_1 is the distance from the object image to the hologram, and X_2 and X_3 are the in-plane displacement vector components. By properly selecting the correct slit aperture one can achieve a sharp change in the fringe visibility necessary to improve the accuracy of X_2 and X_3.

127

Fig.4.15. Interferogram taken using the slit mask of Fig.4.14 of a cantilever beam subjected to bending, showing oscillating fringe visibility

As an illustration, Fig.4.15 shows a reconstructed image of the side surface of a cantilever beam subjected to bending and observed through a slit aperture with a = 80 mm, b = 2 mm, c = 10 mm and m = 6. One can clearly see the regions where the fringe visibility is zero. A remarkable feature of this method is that the fringe visibility becomes modulated when the image is reconstructed and, thus, by properly varying the parameters of the slit aperture, one can broaden the measurement range.

5. Determination of Strain and Stress in an Elastic Body by Holographic Interferometry

The magnitudes of surface displacements in a strained body are, as a rule, negligible compared to its dimensions. This means that the object behaves as a body with a high rigidity, and its strain is linearly related to the displacement. The smallness of the deformation of a body also implies small displacements of its individual points. This allows interpretation of the holograms by means of the theory of elasticity which assumes the deformations to be small and the material to behave elastically. Obviously, this applies to fringe patterns due only to pure deformation of the body without any rigid displacements. The conditions necessary to obtain such interferograms have been discussed in Sect. 2.2.4.

There are three major methods for employing the formalism of the theory of elasticity in strain measurements. These are (i) a direct substitution of the experimentally obtained displacements or strains into the corresponding theoretical expressions, (ii) a comparison of the experimental data with results of calculations based on a mathematical model, and (iii) a combination of calculated and experimental data. A fourth variation, called the methods of analogy, will also be discussed in this chapter, but falls outside this classification. All these methods have been applied but are not as widespread as one might expect given the high potential of present-day holographic interferometry. Most studies are still in explorative or evaluation stages and are limited to simple comparison between experimental and calculated data.

Turning now to specific problems, we begin with studies of plate bending. To evaluate the stresses and strains in this case, one has to know only the deflection distribution and, hence, experiments are confined to measuring one out-of-plane displacement component only. We have mentioned in Chap. 4 the high efficiency with which holographic interferometry can be used in such measurements. The early publications in this area [5.1-3] are of particular interest since they attempted to evaluate the real boundary conditions which are necessary to construct a mathematical model for the calculation of stresses and strains [5.4, 5]. Studies aimed at analyzing effects of the mechanical properties of the plate material itself were reported in [5.6, 7].

The strain gradient on a plane specimen subjected to tension thus permits one to evaluate the structural features, such as holes [5.8] or cracks [5.9-11]. In the latter case one can calculate the various parameters of the linear mechanics of fracture. Studies of the deformation of continuous shells of different shape [5.12-14] have been performed. In addition, several projects have treated specific problems, in particular, near a cutout in a cylindrical shell [5.15] and have determined stresses along the edges of the cutout [5.16, 17]. Studies confined only to measurement of displacement fields on the surface of an object have already been cited in Chaps.3 and 4.

It is not the purpose of this chapter to present a complete and detailed picture of the methods by which holographic information can be used in the solution of numerous problems in solid mechanics. In addition, we do not attempt to cover all possible relationships between the displacements and the deformations on the surface of an object and the nature of its stressed state. These topics need special attention, as it has been done by *Schumann* et al. [5.18].

In this chapter we provide the basic information that is required for a good understanding of how strains and stresses in elastic bodies are determined from experimental data. This should aid the researcher in the solution of concrete problems. To facilitate understanding of the material by specialists in the field of mechanics, we use the accepted notations from the theory of elasticity. In particular, the components of the displacement factor in Cartesian coordinates will be denoted by u, v and w (the u and v components are tangential, and w is normal to the surface of the object under investigation).

5.1 Principal Relations Between Displacements, Strains, and Stresses at a Point

5.1.1 Relations Between the Strain and Displacement Components

The mathematical theory of elasticity considers individual points of a strained object. The position of a point defined by the radius vector **r** with components x, y, z, changes as a result of deformation of the object by the quantity Δ**r** representing a displacement vector **d** with components u, v, w.

The displacement field determines the deformation of the object as a whole and is thus related to its rigidity, while the field of relative displacements or gradients of the vector **d** characterizes the deformed state at individual points of the object, thus providing information on the mechanical behavior and the strength of the material at these points.

The theory of elasticity is based on Cauchy's equations relating the components of deformation and displacement at a point:

• for unit linear strain along the three axes:

$$\epsilon_x = \frac{\partial u}{\partial x}, \quad \epsilon_y = \frac{\partial v}{\partial y}, \quad \epsilon_z = \frac{\partial w}{\partial z} \tag{5.1}$$

• and for the angles of shear in the three planes:

$$\gamma_{xy} = \frac{\partial v}{\partial x} + \frac{\partial u}{\partial y}, \quad \gamma_{yz} = \frac{\partial w}{\partial y} + \frac{\partial v}{\partial z}, \quad \gamma_{xz} = \frac{\partial w}{\partial x} + \frac{\partial u}{\partial z}. \tag{5.2}$$

Besides deformation, the region around a point undergoes a rigid rotation about it. For a small displacement **d**, the angles of rotation about the corresponding axes can be found from

$$\omega_x = \frac{1}{2}\left(\frac{\partial w}{\partial y} - \frac{\partial v}{\partial z}\right), \quad \omega_y = \frac{1}{2}\left(\frac{\partial u}{\partial z} - \frac{\partial w}{\partial x}\right), \quad \omega_z = \frac{1}{2}\left(\frac{\partial v}{\partial x} - \frac{\partial u}{\partial y}\right). \quad (5.3)$$

The gradient of the displacement vector is a second-rank tensor:

$$\underset{\sim}{\mathbf{I}} = \frac{d\mathbf{d}}{d\mathbf{r}} = \begin{bmatrix} \dfrac{\partial u}{\partial x} & \dfrac{\partial v}{\partial x} & \dfrac{\partial w}{\partial x} \\[2mm] \dfrac{\partial u}{\partial y} & \dfrac{\partial v}{\partial y} & \dfrac{\partial w}{\partial y} \\[2mm] \dfrac{\partial u}{\partial z} & \dfrac{\partial v}{\partial z} & \dfrac{\partial w}{\partial z} \end{bmatrix}. \quad (5.4)$$

The above relationships (5.1-3) permit the components of matrix (5.4) to be represented in terms of the components of strains and angles of rotation:

$$\underset{\sim}{\mathbf{I}} = \begin{bmatrix} \epsilon_x & \frac{1}{2}\gamma_{xy}+\omega_z & \frac{1}{2}\gamma_{xz}-\omega_y \\[2mm] \frac{1}{2}\gamma_{yx}-\omega_z & \epsilon_y & \frac{1}{2}\gamma_{yz}+\omega_x \\[2mm] \frac{1}{2}\gamma_{zx}+\omega_y & \frac{1}{2}\gamma_{zy}-\omega_x & \epsilon_z \end{bmatrix}. \quad (5.5)$$

In turn, the matrix (5.5) can be resolved into a symmetrical and a skew-symmetrical matrix:

$$\underset{\sim}{\mathbf{I}} = \begin{bmatrix} \epsilon_x & \frac{1}{2}\gamma_{xy} & \frac{1}{2}\gamma_{xz} \\[2mm] \frac{1}{2}\gamma_{yx} & \epsilon_y & \frac{1}{2}\gamma_{yz} \\[2mm] \frac{1}{2}\gamma_{zx} & \frac{1}{2}\gamma_{zy} & \epsilon_z \end{bmatrix} + \begin{bmatrix} 0 & \omega_z & -\omega_y \\[2mm] -\omega_z & 0 & \omega_x \\[2mm] \omega_y & -\omega_x & 0 \end{bmatrix}. \quad (5.6)$$

Here the first matrix is the well known strain tensor $\underset{\sim}{\mathbf{I}}_\epsilon$, and the second is a tensor characterizing rigid rotation. Since the results of holographic measurements are related to the displacement of points on the surface of an object, one should specifically discuss deformation at these points. Figure 5.1 depicts an arbitrary point on the surface of an object with the z axis normal, and the x and y axes tangential to the surface.

The gradient of the displacement vector $\underset{\sim}{\mathbf{I}}_n$ for a surface element dxdy can be derived from the matrix (5.5 or 6) by deleting the third row which corresponds to the variation of the displacement component along the z axis. If we assume that the z axis coincides with the principal axis of the deformation of a volume element whose face lies on the surface so that the shear angles are equal, $\gamma_{xz} = \gamma_{yz} = 0$, then after deletion of the third row and decomposing the rotation tensor into two terms like (5.6) we find

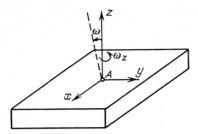

Fig.5.1. Coordinate axes at a point **A** on the object surface, and rotation angle about the z axis ω_z and in the x-y plane, ω

$$\underset{\sim}{\mathbf{I}}_n = \begin{bmatrix} \epsilon_x & \frac{1}{2}\gamma_{xy} & 0 \\ \frac{1}{2}\gamma_{yx} & \epsilon_y & 0 \end{bmatrix} + \begin{bmatrix} 0 & \omega_z & 0 \\ -\omega_z & 0 & 0 \end{bmatrix} + \begin{bmatrix} 0 & 0 & -\omega_y \\ 0 & 0 & \omega_x \end{bmatrix}. \tag{5.7}$$

The first term is the two-dimensional strain tensor $\underset{\sim}{\mathbf{I}}_\epsilon$, the second is the vector of rotation of the element about the normal, and the third, the rotation vector in the x-y plane causing the normal to turn by the angle ω. Both rotations of the normal, ω and ω_z, are shown in Fig.5.1. All components of the tensor $\underset{\sim}{\mathbf{I}}_n$ can be determined from experiment. The components of $\underset{\sim}{\mathbf{I}}_\epsilon$ can be calculated with (5.1,2), while (5.3) can be used to determine ω_z. Recalling that the shear angles γ_{xz} and γ_{yz} are zero, the angles of rotation ω_x and ω_y can be expressed in terms of the derivatives along the axes lying in the plane of the surface element:

$$\omega_x = \frac{\partial w}{\partial y}, \quad \omega_y = -\frac{\partial w}{\partial x}.$$

5.1.2 Relation Between the Stress and Strain Components

The state at a point of a strained object is characterized by the stress tensor

$$\underset{\sim}{\mathbf{I}}_\sigma = \begin{bmatrix} \sigma_x & \tau_{xy} & \tau_{xz} \\ \tau_{yx} & \sigma_y & \tau_{yz} \\ \tau_{zx} & \tau_{zy} & \sigma_z \end{bmatrix}. \tag{5.8}$$

The tensor is made up of normal (σ) and tangential (τ) components acting on the faces of a volume element including the point in question, which are normal to the x, y, and z axes.

The stress components are related to the corresponding strain components through the equations of elasticity

$$\sigma_x = 2G\epsilon_x + \lambda\epsilon_v, \quad \tau_{xy} = G\gamma_{xy},$$
$$\sigma_y = 2G\epsilon_y + \lambda\epsilon_v, \quad \tau_{yz} = G\gamma_{yz},$$
$$\sigma_z = 2G\epsilon_z + \lambda\epsilon_v, \quad \tau_{xz} = G\gamma_{xz}, \tag{5.9}$$

where G is the shear modulus, $\lambda = 2\mu G/(1-2\mu)$ is Lamé's constant, μ is Poisson's coefficient, and $\epsilon_v = \epsilon_x + \epsilon_y + \epsilon_z$ is the first invariant of the strain

132

tensor. In the general case, surface elements will reside in a biaxial stressed state characterized by the stress tensor

$$\underset{\sim}{\mathbf{I}}_\sigma = \begin{bmatrix} \sigma_x & \tau_{xy} \\ \tau_{yx} & \sigma_y \end{bmatrix} .$$

However, in the case of a generalized plane stressed state on a surface typical of thin-walled structures, Eq.(5.9) combined with the condition $\sigma_z = 0$ yields for the unit elongation in the direction of the normal

$$\epsilon_z = - \frac{\lambda\theta}{2G} = - \frac{\mu}{1-\mu} \left(\frac{\partial u}{\partial x} + \frac{\partial v}{\partial y} \right) ,$$

with the strain tensor written in the form

$$\underset{\sim}{\mathbf{I}}_\epsilon = \begin{bmatrix} \epsilon_x & \tfrac{1}{2}\gamma_{xy} & 0 \\ \tfrac{1}{2}\gamma_{yx} & \epsilon_y & 0 \\ 0 & 0 & \epsilon_z \end{bmatrix} .$$

5.2 The Stressed State in Structural Elements

5.2.1 Some Relations from the Approximate Theory of Elasticity

In contrast to the exact mathematical theory of elasticity, the approximate theory permits a substantial broadening of the scope of solvable problems by taking into account only the principal factors governing the deformation of an object, provided the accuracy required in engineering calculations has been assured. This approximation is also applied in the strength analysis of one-dimensional problems. Holographic interferometry may prove to be a useful tool for testing the approximations made, particularly in the specification of boundary conditions.

There are several classical problems in the theory of elasticity, which can be solved by using elementary hypotheses without any recourse to a sophisticated mathematical formalism. Among them are the tension and pure bending of prismatic bars, and the torsion of a circular shaft. Since these problems encompass the main types of deformations, we consider them from a common standpoint where the deformation of an object is described in terms of the displacements occurring on its surface, which, of course, can be measured by holographic interferometry.

Tension of a Straight Bar. When a straight bar is loaded axially, all points in an arbitrary cross section with coordinate x, the outer surface included, displace by an equal amount along the u axis. The unit elongation, in a given cross section, in the longitudinal (ϵ_x) and transverse (ϵ_y) directions depends on the axial derivative of the displacement

$$\epsilon_x = \frac{\partial u}{\partial x}\,, \quad \epsilon_y = -\,\mu\,\frac{\partial u}{\partial x}\,.$$

For uniaxial tension the normal stresses σ_x will be

$$\sigma_x = E\,\frac{\partial u}{\partial x}$$

by Hooke's law. One can now express the axial load of the bar as

$$N_x = \sigma_x F$$

where F is the cross sectional area.

Torsion of a Circular Bar. The cross sections of a circular bar subjected to torsion rotate rigidly about its x-axis and all points in the cross section, including the bar's surface, undergo in-plane displacements. If we consider surface points lying on the same generatrix perpendicular to the y axis, then their displacement can be denoted by the v component.

The axial derivative of this displacement determines the main parameters characterizing the deformation of a bar:
• shear angle: $\gamma_{xy} = \partial v/\partial x$,
• unit angle of twist: $\theta_x = (1/r)(\partial v/\partial x)$, where r is the bar radius.

Knowing the derivative $\partial v/\partial x$ one can readily determine:
• the angle of rotation of the bar's cross section on a length ℓ:

$$\varphi = \frac{1}{r}\int_0^{\ell} \frac{\partial v}{\partial x}\,dx\,,$$

• the maximum in-plane stress

$$\tau_{max} = G\,\frac{\partial v}{\partial x}$$

• the bending moment

$$M_x = GJ_p\theta_x$$

where J_p is the polar moment of inertia of the cross section.

Pure Bending of a Prismatic Bar. Pure bending deformation occurs if the bar is under the action of a bending moment only and, in theory, a plane section before bending remains a plane after bending. The deformation of a rectangular bar under bending is illustrated by Fig.5.2.

The deflection of an arbitrary point of the bar can be found from the following expression

134

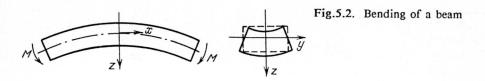

Fig.5.2. Bending of a beam

$$w = \frac{1}{2r_x} [x^2 + \mu (z^2 - y^2)] , \tag{5.10}$$

where r_x is the radius of curvature of the neutral axis.

The equation of the elastic axis is that of a parabola

$$w = \frac{x^2}{2r_x} ,$$

which, for small deflections, approximates to an arc of radius r_x. The curvature in the transverse direction is an odd function, i.e.,

$$\frac{1}{r_y} = - \frac{\mu}{r_x}$$

This also implies that any surface with z = const., the outer surface included, has an anticlastic shape. This fact is utilized for the determination of Poisson's coefficient, and can be done by holographic interferometry as well as other means (Sect.4.1.1).

The second derivative of the deflection, which yields the curvature, allows determination of the unit linear strain and stress, in particular, on the outer surface of a bar of height h

$$\epsilon_x = \frac{h}{2} \frac{\partial^2 w}{\partial x^2} , \quad \sigma_x = E \frac{h}{2} \frac{\partial^2 w}{\partial x^2} \tag{5.11}$$

as well as the bending moment

$$M_y = E J_y \frac{\partial^2 w}{\partial x^2} ,$$

where J_y is the moment of inertia of a cross section with respect to the neutral axis.

Uniaxial Tension of a Plate with a Circular Hole. We now consider two-dimensional problems in the theory of elasticity to the solution of which one can also use holographic interferometry data.

Figure 5.3 illustrates the loading of a plate of uniform thickness with a circular hole where stress concentration is observed (Kirsch's problem). In the vicinity of the hole, a biaxial stressed state with the corresponding

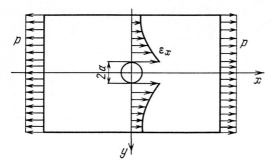

Fig.5.3. Plate of uniform thickness with a circular hole, under tensile stress

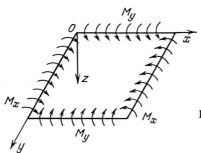

Fig.5.4. Pure bending of a thin plate

strain components prevails. The strains ϵ_x are affected particularly strongly along the y axis passing through the center of the hole, and are described by

$$\epsilon_x = \epsilon_{x0} \left[1 + \frac{a^2}{y^2} \left(\frac{1-3\mu}{2} \right) + \frac{a^4}{y^4} \frac{3(1+\mu)}{2} \right] \qquad (5.12)$$

where ϵ_{x0} = p/E is the strain in a region sufficiently far away from the hole. The expression in brackets presents the strain concentration $K_\epsilon = \epsilon_x / \epsilon_{x0}$ whose maximum value at the point y = a is 3. The values of ϵ_x given by (5.12) are shown graphically in Fig.5.3 for a symmetrical problem. The unit strain along the other axis, ϵ_y, is much less than ϵ_x and is therefore neglected. The efficiency of holographic interferometry in the solution of this problem has been demonstrated by *Borynyak* et al. [5.19].

Pure Bending of a Thin Plate. The loading of a thin plate by moments distributed uniformly along its edges is illustrated in Fig.5.4 where the xy plane is defined to be the middle (neutral) surface of the plate before its deformation [5.20]. The moments M_x and M_y are aligned parallel to the y and x axis, respectively. The xz and yz planes are parallel to the principal planes of the curvature whose values at a given point on the neutral surface are

$$\frac{1}{r_x} \doteq - \frac{\partial^2 w}{\partial x^2}, \quad \frac{1}{r_y} = - \frac{\partial^2 w}{\partial y^2}. \qquad (5.13)$$

136

The unit elongations of an element on the outer surface of a plate of thickness h in the directions of the corresponding axes can be determined from the expressions

$$\epsilon_x = \frac{h}{2} \frac{\partial^2 w}{\partial x^2}, \quad \epsilon_y = \frac{h}{2} \frac{\partial^2 w}{\partial y^2}.$$

Using (5.9) for a plane problem, one can now calculate normal stresses on the surface

$$\sigma_x = -\frac{E}{1-\mu^2} \frac{h}{2} \left[\frac{\partial^2 w}{\partial x^2} + \mu \frac{\partial^2 w}{\partial y^2} \right],$$

$$\sigma_y = -\frac{E}{1-\mu^2} \frac{h}{2} \left[\frac{\partial^2 w}{\partial y^2} + \mu \frac{\partial^2 w}{\partial x^2} \right].$$

The bending moments are also determined in terms of the derivatives of deflections

$$M_x = -D \left[\frac{\partial^2 w}{\partial x^2} + \mu \frac{\partial^2 w}{\partial y^2} \right],$$

$$M_y = -D \left[\frac{\partial^2 w}{\partial y^2} + \mu \frac{\partial^2 w}{\partial x^2} \right],$$

(5.14)

where $D = Eh^3/[12(1-\mu^2)]$ is the flexural rigidity of a plate.

Consider some specific cases of pure bending of plates:

(i) The bending moments are distributed uniformly along the edges of a plate, i.e., $M_x = M_y = M$. For the curvature in both directions we have

$$\frac{1}{r_x} = \frac{1}{r_y} = \frac{M}{D(1+\mu)},$$

and the plate bends to a spherical surface. However, integration of the expressions for the second derivatives yields an equation of a paraboloid of revolution:

$$w = -\frac{M}{2D(1+\mu)}(x^2 + y^2).$$

This inconsistency can be accounted for by the approximate nature of the expression for the curvature in (5.13) in terms of the second derivative.

(ii) The bending moments are equal in absolute magnitude and opposite in sign, $M_x = -M_y$. In this case the curvatures will have opposite signs but will also be equal in absolute magnitude: $1/r_x = -1/r_y$, the deflection being described by an anticlastic surface whose equation takes on the form

$$w = \frac{M(x^2 - y^2)}{2D(1 + \mu)} \, .$$

(iii) Bending to a cylindrical surface where the curvature is

$$1/r_y = \partial^2 w / \partial y^2 = 0 \, .$$

As follows from (5.14), both bending moments should appear in this case. If only one moment M_x is present, the plate will acquire an anticlastic shape. For a thin rectangular plate, however, the expressions derived for pure bending of a bar will be sufficiently accurate.

Calculation of the bending of circular plates of uniform thickness is usually done in polar coordinates where the expressions for the radial (σ_r) and tangential (σ_θ) stresses can be written as

$$\sigma_r = - \frac{E}{1-\mu^2} \frac{h}{2} \left[\frac{\partial^2 w}{\partial r^2} + \frac{\mu}{r} \frac{\partial w}{\partial r} \right] ,$$

$$\tag{5.15}$$

$$\sigma_\theta = - \frac{E}{1-\mu^2} \frac{h}{2} \left[\frac{1}{r} \frac{\partial w}{\partial r} + \mu \frac{\partial^2 w}{\partial r^2} \right] .$$

We now consider the application of holographic interferometry to the calculation of stresses in a circular plate with a hole, where the plate is clamped at the outer edges and the edges of the hole are supported elastically. Figure 5.5 shows a two-exposure fringe pattern obtained on such a plate made of D16 alloy with the diameters $D_1 = 8$ mm and $D_2 = 70$ mm, and h = 2 mm thick, for a loading regime where the spring of the central support has a high stiffness. A spline fitting performed on the experimen-

Fig.5.5. Interferogram of a strained circular plate with a hole clamped along the outer and inner edges

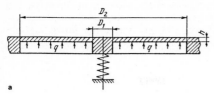

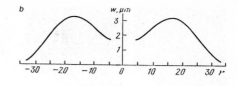

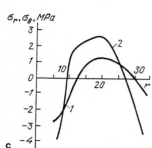

Fig.5.6. (a) Loading of a plate with a hole, (b) deflection curve from this loading, and (c) distribution of (*1*) radial σ_r and (*2*) tangential stress σ_θ

tally obtained displacements yielded the radial dependence of the deflection and of its derivatives. The character of the plate's deformation shown schematically in Fig.5.6a can best be visualized by means of the deflection curve presented in Fig.5.6b. Figure 5.6c displays graphically the variation of the radial (σ_r) and tangential (σ_θ) stresses for the outer surface of the plate calculated by (5.15).

Stress calculation for points on the free surface of the plate, in particular, along the edges of the hole, is greatly simplified since at these points the uniaxial stressed state prevails. The stresses tangential to the edge at these points can be calculated by

$$\sigma_\varphi = E\frac{h}{2}\left[\frac{1}{r^2}\frac{\partial^2 w}{\partial\varphi^2} + \frac{1}{r}\frac{\partial w}{\partial r}\right]. \tag{5.16}$$

Figure 5.7 shows a fringe pattern for a region near the hole obtained with the plate clamped along the edges and subjected to the action of oppositely directed force P. The points of application of the forces are clearly indicated in the interferogram. The fringe pattern was obtained by the method discussed in Sect.2.3.3. The position of the bright zero motion fringe specifies the boundary where the deflections reverse their sign. The radial direction of the fringes along the edges of the hole allows us to neglect the second term in the parentheses in (5.16).

The distribution of σ_φ stresses along the edges of a 50 mm diameter circular hole, presented in Fig.5.8, was obtained on a plate of D16 alloy h = 2 mm thick for P = 2.6 N.

Tension of a Cylindrical Shell with a Hole. Measurements of the u, v, w displacements in a cylindrical coordinate system can be directly used, just as this is done in the preceding example, for stress determination on

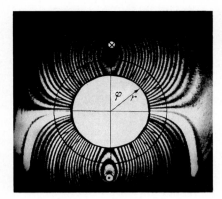

Fig.5.7. Interferogram of plate deflection near a circular hole (the point and the cross specify the points where the forces are applied)

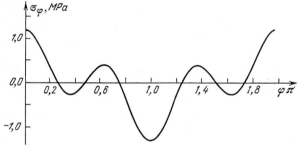

Fig.5.8. Distribution of the stress σ_φ along the edge of a circular hole

the surface of a shell along the edges of a hole [5.21]. The corresponding tangential stresses can be expressed in the following way as a function of angle φ:

$$\sigma_\varphi = E\left[\frac{1}{r}\frac{\partial v_1}{\partial \varphi} + \frac{u_1}{r} + \frac{\cos^2\varphi}{R}w\right], \qquad (5.17)$$

where R is the shell radius, and r is the hole radius. The displacement components u_1 and v_1, normal and tangential to the edge, respectively, are determined from

$$u_1 + iv_1 = (u + iv)\exp(-i\varphi) .$$

The interferometric data presented in Fig.3.6 can be used to find the stresses σ_φ appearing in a shell under tension. The expressions derived for the displacement components u and v can be fitted by trigonometric series of the type

$$u = A(1) + \sum_{k=2}^{n} A(k)\cos[(k-1)\varphi] ,$$

140

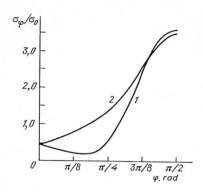

Fig.5.9. Stress concentration along the edge of a hole obtained by (1) holographic interferometry, and (2) finite element calculation

$$v = B(1) + \sum_{k=2}^{n} B(k)\sin[(k-1)\varphi] \ .$$

The expression for the w component can be deduced from experimental data by means of an interpolation spline. Curve 1 in Fig.5.9 gives an idea of the concentration of σ_φ stresses along the outer edges of the hole relative to the σ_0 stresses at the face ends of the cylinder. Curve 2 in the same figure displays a similar relationship derived from finite-element calculations [5.22].

5.2.2 Numerical-Experimental Methods

The present calculational methods used in solid mechanics and based on finite-difference, finite-element, and boundary-element techniques provide the possibility of studying the stressed state in structural components of complex geometrical shape. Numerical methods, however, can often yield inaccurate results because of an incomplete and only approximate knowledge of the boundary conditions. This difficulty can be alleviated by using supplemental experimental information which reflects the real conditions under which an object is stressed. Particularly promising approaches are those where experimental data are used to formulate the boundary conditions while the numerical calculations are employed to analyse the stresses and strains throughout the object under investigation.

The first publications that reported on the use of semi-empirical methods involving the finite-difference methods are [5.23-25]. If a boundary value problem is formulated in terms of displacements, then one has to describe the displacement vector components u, v, w over the entire boundary of the region to be studied. The inner point displacements which can be used to calculate the stresses and strains by means of well-known relations are found by numerically solving Lamé's equations

$$\Delta u + \frac{1}{1-2\mu}\frac{\partial \epsilon_v}{\partial x} = 0 \ ,$$

141

$$\Delta v + \frac{1}{1-2\mu} \frac{\partial \epsilon_v}{\partial y} = 0 \, , \qquad\qquad\qquad (5.18)$$

$$\Delta w + \frac{1}{1-2\mu} \frac{\partial \epsilon_v}{\partial y} = 0 \, ,$$

where $\epsilon_v = \partial u/\partial x + \partial v/\partial y + \partial w/\partial z$ is the dilational strain satisfying the equation $\Delta \epsilon_v = 0$ for zero or constant body forces. However, even in cases where holographic interferometry is used it is not always feasible to perform measurements over the entire boundary surface and, hence, to formulate the boundary value problem in a good way. Special methods of numerical determination of the boundary conditions which cannot be found from experiment, have been proposed [5.26].

The cases of a bar with a circular hole subjected to tension [5.27-28] and of bending of an I-beam with a hole in the wall [5.29] convincingly demonstrate that combined use of experimental information and of the finite-element method can be efficient in stress determination in regions with high stress gradients. It permits solving problems which are otherwise difficult or impossible to investigate by experimental means only. This combined approach also promises a substantial reduction in the volume of the required experimental work.

Holographic interferometry was combined with the finite-element method to study the tension of a cylindrical shell with a circular hole (Fig. 5.10) [5.22]. The symmetry of the problem allows one fourth of the shell to be considered. The finite-element mesh model is depicted in Fig. 5.11. Triangular elements responding to both membrane and bending loads were used [5.30]. The first stage of the work was to test the operation of the elements in the vicinity of the circular hole and to evaluate the chosen degree of discretization. The load was a unit tensile stress along the x axis applied along the FJ line (Fig.5.11). Depending on the actual conditions of the shell loading, the displacement-vector components along the y and z axes on the same line as well as the angles of rotations of the cross sections about these axes were all taken to be zero. The line AL was clamped at nodes in the direction of the x axis. By symmetry, the angles of rotation of the cross sections on this line about the y and z axes are considered to be zero, as are the displacement-vector components along the y axis, and the angles of

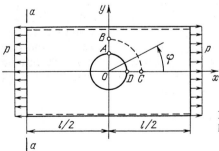

Fig.5.10. Cylindrical shell with a circular hole subjected to a tensile stress

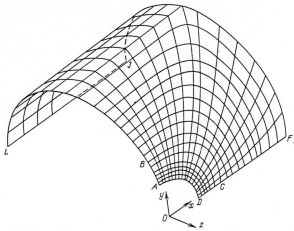

Fig.5.11. Mesh for one fourth of the shell of Fig.5.10

rotation of the cross sections about the x axis along the lines LJ and DF. The distributions of the membrane (σ_M) and bending (σ_B) normal stresses along the edge of the hole AD are in agreement with the well-known analytical solutions for a similar infinite shell, the maximal value of σ_M being comparable with the experimental figure obtained by the method involving optically sensitive coatings. Thus, the type of the finite element chosen and the degree of discretization of the object provide the required accuracy for determining stress concentration by means of the finite-element method.

In studies of the stressed state along the edges of a hole by numerical-experimental techniques one can use a fragment of the finite-element mesh bounded by the ABCD contour (Fig.5.11). The three components of the displacement vectors at the nodes of the coordinate mesh on the specimen, which was adequate to a finite-element division, were determined by holographic interferometry. The distribution of the displacement components along the edge of the hole AD is shown in Fig.3.6 for a corresponding tensile stress, with similar relations obtained for the region BC of the boundary contour.

When prescribing boundary conditions for the fragment ABCD of the shell, the amount of the required experimental information can be reduced by specifying the three components of the displacement vectors at the nodes only along the boundary line BC. The kinematic boundary conditions used in the finite-element calculation of one fourth of the shell can be employed along the DC line. Displacements along the AD line are not specified. Along the AB line one prescribes the component of the u displacement, since in the tension of a real specimen clamped in the a-a cross section (Fig.5.10) the points of this component move various distances along the x axis. The membrane (σ_M) and bending (σ_B) stresses along the AD line of the hole obtained by combined calculation and experiment, and normalized to p = 1, are illustrated in Fig.5.12 with the curves 3 and 4, respectively. They coincide with the results of a finite-element calculation performed on

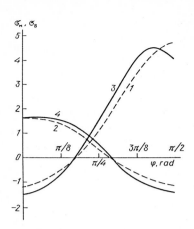

Fig.5.12. Distribution of (*1,3*) membrane and (*2,4*) bending stress on the outer surface of the shell along the edge of a hole. Dashed curve - finite-element calculation, solid curve - numerical-experimental data

one fourth of the shell (curves *1* and *2* in Fig.5.12) along almost the entire AD line.

The efficiency of the boundary-element methods, also called the boundary integral-equation methods, is accounted for by the fact that, in contrast to the finite-element approach, only the boundary of the object under investigation is divided into elements [5.31]. In this way one less dimension of the problem needs to be considered, thus reducing the time needed for the preparation of the basic data. In addition, the boundary-element method permits study of areas where concentrated loads are applied, as well as regions of infinite dimensions. It may prove convenient for evaluating experimental data on the surface deformation of objects, particularly those obtained by holographic interferometry, which provide a complete picture of the deformation [5.32, 33]. In some cases the combination of the boundary-element method with holographic interferometry allows simplification of experimentally derived boundary-value conditions [5.34].

5.3 Methods of Analogies

Methods of analogies are essentially based on the observation that phenomena of different nature are sometimes described by the same mathematical relationships. The same stress and deflection functions are used in solid mechanics as in the bending of membranes or plates. The latter, as already mentioned, can conveniently be determined by means of holographic interferometry. We consider two kinds of problems where analogies are employed.

5.3.1 Prandtl's Membrane Analogy

This analogy is used in studying the torsion of prismatic bars [5.35]. The approach of the theory of elasticity to these problems is based on the assumption of a rigid rotation of the cross sections about the bar axis and their distortion along the axis. The rotation is parameterized by the angle of torsion per unit length θ_x. This condition is defined by the torsion function $\varphi(y, z)$

$$u = \theta_x \varphi(y, z) \ .$$

This function satisfies the harmonic equation and determines the stressed state of all points in a cross section under torsion. To facilitate the analysis of the boundary conditions, the torsion function φ can be replaced by a conjugate function Ψ which is likewise harmonic

$$\frac{\partial \varphi}{\partial y} = - \frac{\partial \Psi}{\partial z} , \frac{\partial \varphi}{\partial z} = \frac{\partial \Psi}{\partial y} \ .$$

Along the boundary of the cross section, Ψ is specified by the values of the function itself, rather than by those of its derivative as is the case with the function φ.

The most convenient form, however, to calculate the stressed state over a cross section is Prandtl's stress function

$$F(y, z) = G\theta_x \left[\Psi - \frac{y^2 + z^2}{2} \right] \ .$$

In a cross section of the bar this function satisfies Poisson's differential equation

$$\Delta F = \frac{\partial^2 F}{\partial y^2} + \frac{\partial^2 F}{\partial z^2} = - 2G\theta_x \ . \tag{5.19}$$

The main advantage in introducing the function F is that its value at the edges of a cross section can be taken to be zero. The tangential stress components can be expressed in terms of the derivatives of this function

$$\tau_{xy} = - \frac{\partial F}{\partial z} , \quad \tau_{xz} = \frac{\partial F}{\partial y} \ ,$$

while the torsional moment can be represented as an integral over the area of the cross section

$$M_x = 2 \iint F \, dy \, dz \ . \tag{5.20}$$

Prandtl pointed out a mathematical analogy between the stress function F and the deflection w of a membrane subjected to a transverse uniformly distributed load q and simultaneously stretched with the uniform tension p. From the equilibrium of an infinitesimal element of the membrane in the direction of the z axis (Fig. 5.13) it follows that

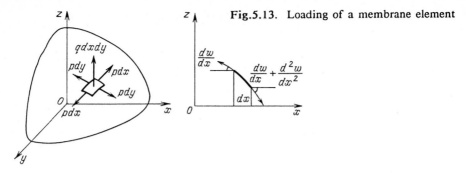

Fig.5.13. Loading of a membrane element

$$qdxdy + p\left[\frac{\partial^2 w}{\partial x^2} + \frac{\partial^2 w}{\partial y^2}\right]dxdy = 0 \ ,$$

and

$$\Delta w = \frac{\partial^2 w}{\partial x^2} + \frac{\partial^2 w}{\partial y^2} = -\frac{q}{p} \ . \tag{5.21}$$

One readily sees that the expressions for the function F and deflection w will coincide provided the right-hand sides of (5.19,20) are equal, as will the boundary conditions if the membrane is drawn onto a rigid loop of the same shape as the contour of the cross section of the prism. In the general case, the relation between the stress function and membrane deflection can be written as

$$F = 2G\theta_x \frac{p}{q} w \ . \tag{5.22}$$

Thus, the elastic surface of a membrane reproduces the function F on some conventional scale. By intersecting this surface with planes where w = const., we obtain contour lines corresponding to F = const.. The stress vectors are tangential to these lines, and their density indicates the magnitude of stress. In this way, one can also determine the bending moment which, by (5.20), should be equal to twice the volume bounded by the surface of the deformed membrane.

By using holographic interferometry with the optical arrangements discussed in Sect.4.1.1, one can obtain patterns of equal deflection fringes. *Robertson* et al. [5.36] presented a comprehensive analysis of the application of holographic interferometry to the membrane analogy. They used a 50 μm thick, pressure-loaded rubber membrane fitted closely to a rigid plate with cutouts of the same shape as the cross sections under study. One of the circular cutouts, of radius r_0, served to determine the q/p ratio in the shell through the following relation

$$\frac{q}{p} = \frac{4}{r_0^2} w_{max} \ .$$

5.3.2 Analogy Between the Plane Stressed State and Bending of a Plate.

Airy's stress function Φ used in plane problems of the theory of elasticity satisfies the biharmonic equation [5.35]

$$\Delta\Delta\Phi = 0 . \qquad (5.23)$$

The stress components can be defined in terms of the derivatives of this function:

$$\sigma_x = \frac{\partial^2 \Phi}{\partial y^2} , \quad \sigma_y = \frac{\partial^2 \Phi}{\partial x^2} , \quad \tau_{xy} = -\frac{\partial^2 \Phi}{\partial x \partial y} .$$

The same differential equation can be used to describe the deflection of a plate loaded only along the edges [5.20]

$$\Delta\Delta W = 0 . \qquad (5.24)$$

The mathematical analogy between (5.23 and 24) permits one to consider the stress function Φ as an elastic surface of deflection of the same plate provided that the boundary conditions are also analogous. The stress function can be expressed in terms of the plate deflection using a numerical parameter C:

$$\Phi = Cw . \qquad (5.25)$$

In the problem of lateral plate bending, the analogs to the stresses will be curvatures and relative torsions of the surface about the x and y axes at the same points defined in terms of the second derivatives of the deflections:

$$\sigma_x = \frac{C}{r_y} = C\frac{\partial^2 w}{\partial y^2} , \quad \sigma_y = \frac{C}{r_x} = C\frac{\partial^2 w}{\partial x^2} , \quad \tau_{xy} = \frac{C}{r_{xy}} = C\frac{\partial^2 w}{\partial x \partial y} . \qquad (5.26)$$

For this analogy to hold, the crucial problem is to provide the necessary boundary conditions. It was found that the static boundary conditions of one problem turn out to be analogous to the kinematic boundary conditions of the other, and vice versa. Indeed, within the straight-line section S of the boundary of a plate under tension subjected to a normal pressure p the stress function satisfies the condition

$$\Phi_S = \int(\int p dS) dS . \qquad (5.27)$$

The normal derivative of Airy's function at the boundary will then be zero:

$$\left.\frac{\partial\Phi}{\partial n}\right|_S = 0 . \qquad (5.28)$$

147

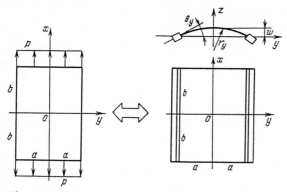

Fig.5.14. Similar boundary conditions for a plate under tensile and bending stress

One can use (5.25) to obtain, for the static boundary conditions (5.27,28) of the stress function, kinematic boundary conditions for the same region of the plate under lateral bending:

$$Cw_S = \int(\int pdS)dS \ , \quad C\frac{\partial w}{\partial n}\bigg|_S = 0 \ . \tag{5.29}$$

As an illustration, we consider uniaxial tension of a rectangular plate subjected to a uniform lateral pressure p. As follows from (5.29):
• the constant-curvature deflections of radius $r_y = C/p$, without tilting of the normals under bending, should correspond to the loaded faces of a plate in tension;
• rigid clamping under bending turned by an angle $\theta_y = a/r_y$ should correspond to the free side faces of a plate in tension.
The boundary conditions for a plate subjected to tension and bending are depicted in Fig.5.14. From the pattern of deflections under bending one can derive the stresses at any point of the plate in tension using (5.26). A two-exposure holographic interferogram, used to determine the deflection field for a plate subjected to bending, was used to find the stresses in a plate with side holes under tension [5.37].

One should bear in mind that the use of the membrane analogy has technical difficulties associated with the implementation of the boundary conditions, and requires double differentiation for the stress calculation, an operation of inherently poor accuracy.

5.4 Methods for Determining Derivatives of Displacements

As already mentioned, evaluation of the stressed state at various points on the surface of a body requires calculation of the derivatives of the displacement function. However, differentiation is known to be highly sensitive to even small errors in the initial data; this may bring about substantial errors in the final results. In the present section, we consider possible ways of reducing this problem in holographic interferometry by using numerical and optical methods.

5.4.1 Numerical Methods

Determination of derivatives in the analysis of experimental data, in particular, in problems solvable by holographic interferometry, has been a subject of considerable interest [5.38-40]. No concrete recommendations or a common approach to the problem have been developed up to now, however, simply because the problem is mathematically ill-defined.

In selecting a method of numerical differentiation one should be guided by preliminary information on the character of the deformation of the object under study. This information can be derived from a physical analysis of the assumed character of deformation or from a preliminary experiment. Of course, the choice of the method of differentiation and of the degree of allowable error is also related to the particular solution of the given problem. Here we consider some aspects of the widely used numerical methods for differentiation. These are based on approximate values of a function at discrete points, namely, the least-squares, finite-difference, and spline methods.

In the least-squares method, the function is first approximated by an analytical relationship which is usually represented by a polynomial throughout the region in question. The coefficients of the polynomial are found by minimizing the sum of the squared deviations of all points, i.e. $\Sigma_{i=1}^{i=m}(f_i - y_i)^2$. Here y_i and f_i are the ordinates of the measured point and of the approximating polynomial, respectively. By optimizing the degree of the polynomial one can improve the accuracy of the differentiation. The degree of the polynomial, however, does not typically exceed 5 or 6, which is due to a progressive degradation of the matrix for a system of normal equations. If the degree of the approximating polynomial (or the corresponding parameters of any other fitting function) have been appropriately chosen, then the least-squares method is reasonably insensitive to random experimental errors. This method is particularly good in cases where one can take the general analytical solution to the problem as the approximating function. Thus, in the treatment of experimental data on the bending of a plate or beam, for instance, it is appropriate to use the equation for their elastic axis.

In cases where the chosen approximating function cannot account for all the features of the applied strain field, the use of the least-squares method is inappropriate. Indeed, under these conditions, local features in the region of strain concentration will be unjustifiably smoothed out, and the concentration effect strongly underestimated. Increasing the degree of the approximating polynomial will, unfortunately, not solve this problem but lead to the appearance of spurious peaks which reflect a random distribution of deviations in the experimental data.

The finite-difference method makes implicit use of a piecewise-polynomial approximation of the displacement relation and is fairly sensitive to

149

measurement errors. Its application is thus limited to finding second-order derivatives. In analysing experimental data one usually employs numerical differentiation and the relationships are expressed directly in terms of experimental data. Linear polynomials permit calculation only of the first derivatives at both nodal points of the interval h by the expression

$$y_i' = y_{i-1}' = \frac{y_i - y_{i-1}}{h} .$$

Quadratic polynomials are used to calculate the first derivative at the middle and end points in two adjacent intervals of equal width h

$$y_i' = \frac{1}{2h}(y_{i+1} - y_{i-1}) , \quad y_{i-1}' = \frac{1}{2h}(y_{i+1} - 4y_i + 3y_{i-1}) ,$$

as well as to derive an expression for the calculation of the second derivatives which are equal at all nodes

$$y_{i-1}'' = y_i'' = y_{i+1}'' = \frac{1}{h^2}(y_{i-1} - 2y_i + y_{i+1}) .$$

As seen from the above expressions, the error due to differentiation depends on the magnitude of the step h. Therefore, the selection of the step or mesh size in the finite-difference method becomes of crucial importance. One criterion for the selection of nodal points that can be used is the fact that this method permits obtaining more than one value of the derivative at the same point. These values will be close to one another when the mesh size is close to optimal. As a consequence, the errors of the derivatives will also be minimal. While this criterion is not rigorous, it enables one to improve the accuracy of the differentiation and to evaluate the errors.

Among the numerical techniques the spline method has certain advantages. In spline approximation, the experimental function is replaced by a piecewise-analytical relation S(x) with ℓ continuous first derivatives and discontinuities at the nodal points of the $\ell+1$ derivative [5.41]. This method may, in a certain sense, be considered as intermediate between the above-mentioned methods since in constructing a spline one simultaneously implements the properties of integrity and piecewise character of the approximating function.

The cubic spline which has continuous first and second derivatives is presently in widespread use in the treatment of experimental data. Throughout the domain divided by m nodes into m-1 intervals, the spline function consists of m-1 nonoverlapping cubic polynomials $S_i(x)$. For each interval one can write the condition of linearity for the second derivative of

$$S_i''(x) = \frac{x_i - x}{h_i}S_{i-1}'' + \frac{x - x_{i-1}}{h_i}S_i'' . \qquad (5.30)$$

The expressions for the first derivative $S_i'(x)$ and the function proper $S_i(x)$ can be obtained by consecutive integration of (5.30). By virtue of the continuity of the first derivative $S_i'(x)$, its values derived from expressions for the adjacent intervals are assumed to be equal. As a result, one can write as many relationships as there are intermediate nodal points and thus obtain a system of $m-2$ equations of the form

$$S_{i-1}'' h_i + 2S_i'' (h_i + h_{i+1}) + S_{i+1}'' h_{i+1} = 6 \left(\frac{y_{i+1} - y_i}{h_{i+1}} - \frac{y_i - y_{i-1}}{h_i} \right) . \qquad (5.31)$$

The system of (5.31) has m unknown values of the second derivatives S_i'' and can be solved if two more relationships are provided. These relationships are usually derived from physical assumptions about the behavior of the function $S_i(x)$ at the boundary of the region under study. This accounts for a poor accuracy of the spline method near the boundary. Note that the set (5.31) is analogous to the equations of three moments for the displacement of supports, which, in structural mechanics, describe the absence of breaks in a beam above intermediate supports. This cubic spline function can be used to develop a mechanical interpretation by using the analogy to the deformation of a beam resting on sliding supports.

In addition to the interpolation spline whose values at the nodes coincide with the given points on the surface, one can also consider variants of a smoothing spline, which allows deviation from these points. A mechanical analog of such a spline is a multispan beam on elastic supports which are shifted with respect to the original position. While the approximating spline permits improvement in accuracy of the derivative by some smoothing of experimental errors, it does require additional data on the rigidity of each support w_i and on the beam rigidity ξ (i.e., of the spline itself). Beam deflections above the supports, that is, the values of the spline at nodal points, can be derived by minimization of the deviations from the given points Δy_i with a simultaneous minimization of the curvature S_i''

$$E = \sum_i w_i (\Delta y_i)^2 + \xi \int [s_i'' (x)]^2 dx . \qquad (5.32)$$

As is the case with all numerical methods, the choice of the mesh size is of crucial importance for the spline technique, too. This relates to both interpolation and smoothing splines; in the latter case, however, the error due to differentiation increases more slowly with increasing number of nodes, provided that the smoothing parameter ξ is chosen appropriately.

The effect of the number of nodes on the deflection derivative can be illustrated by the case of pure bending of a cantilever beam. Spline interpolation was performed with different numbers of experimental points obtained from the same holographic interferogram. The measurement error in the ordinate was 0.1 fringe. Figure 5.15a shows the deflections expressed in terms of the fringe number n (curve 1) and the first derivatives (curve 2)

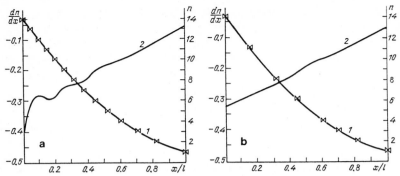

Fig.5.15. (*1*) Deflection and (*2*) first derivative functions for (a) 14 and (b) 8 nodal points

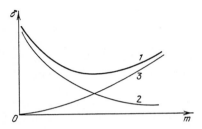

Fig.5.16. Error of derivative determinations vs. number of nodes: (*1*) total, (*2*) errors due to model, and (*3*) experimental errors

within a section of the beam of length ℓ obtained with 14 nodal points. Figure 5.15b depicts similar curves constructed with 8 nodal points. The deflection function is seen to be reproduced well in both cases but it is obvious that the behavior of the first derivative with 14 nodes does not reflect the real conditions of beam deformation but rather random errors in the determination of fringe coordinates.

In all these methods, the accuracy of differentiation depends essentially on the number of nodal points and their location. Finding this relationship, however, is a complex and largely unsolved problem in the theory of random processes [5.38]. The dependence of the error in differentiation, δ, on the number of nodes, m, for a given measurement accuracy is shown graphically in Fig.5.16 (curve *1*). The curve passes through a minimum which corresponds to the minimum number of nodal points necessary to best solve the problem in question. This character of curve *1* is accounted for by a simultaneous action of two opposite factors. On the one hand, increasing the number of nodes reduces the error in approximating the derivative with accurately given initial data (curve *2*), while on the other hand, the error of differentiation increases when the mesh becomes finer because of the effect of random experimental errors (curve *3*). Improving the measurement accuracy will lower curve *3* which will shift the minimum of curve *1* down and to the right. Thus, by making the mesh intervals smaller, one is then able, in this case, to improve the accuracy of differentiation.

Dändliker et al. [5.42] reported an error in fringe-order measurement of 10^{-3}, which permitted determination of the maximum curvature, i.e., of the second derivative, of the cantilever beam deflection with an error of about 2.5% [5.43].

5.4.2 Optical Methods

By making proper use of the various parameters of holographic interference fringes (localization, visibility, spacing, etc.) one can, in some cases, directly determine the derivatives of the displacement-vector components. Indeed, the fringe-localization equations (2.74,75) contain derivatives of the displacement-vector components which can thus be deduced from fringe-localization measurements [5.18,44]. Linear deformations on the surface of an object can be found from studies of fringe contrast [5.45,46]. The derivatives can be determined by using the so-called fringe vector introduced by *Stetson* [5.47,48] or by the fringe-interpretation method where fringe spacing is the input quantity (Sect.2.3.6). To measure the deformation at individual points of an object surface one can use an optical set-up with two illuminating beams [5.49]. All these methods have not yet been used widely because of considerable difficulties associated with the measurement of fringe localization and contrast, the complexity of the equipment involved, and the low measurement precision. However, as the experimental techniques keep improvng and, which is most essential, the measurements become ever more automated, it is expected that this situation will change.

In cases where the fringe pattern can be interpreted as the locus of points with equal displacements, the derivatives can be determined by the well known moiré-pattern method [5.50,51]. It is based essentially on measuring the shift of two identical fringe patterns with respect to one another. Fringe patterns can be shifted optically by defocussing the picture [5.52]. This approach is the easiest to implement in sandwich holography [5.53]. The moiré fringes thus obtained yield information on the derivative of the displacement-vector component recorded on the original interferogram in the direction of the shift. If one can obtain the field of the tangential displacement vector components in a pure form, then the moiré technique can help in determining linear deformations. This method is most widely used in plate-deflection problems, for which, as was shown in Sect.4.1.1, the deflection field can be measured relatively simply.

If the relative shift occurred in the x direction by an amount Δx, one can write the following equation for the moiré fringes

$$\frac{\partial w}{\partial x} = \frac{(n_m + \frac{1}{2})\lambda}{2\Delta x} \tag{5.33}$$

where n_m is the moiré fringe order, and λ is the wavelength of light. The derivation of (5.33) can be found in [5.54] by *Kozachok*. As an illustration, Fig.5.17 shows a moiré pattern observed on a square plate clamped along the edges and subjected to a distributed load.

Another method of determining the derivative of a deflection function involves examining the fringe pattern obtained by a superposition of the

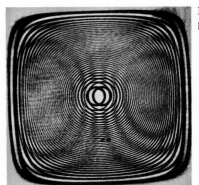

Fig.5.17. Moiré pattern obtained for a square plate under bending stress

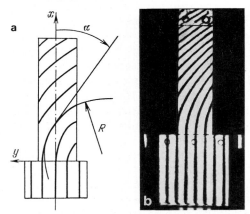

Fig.5.18. (a) Calculated fringe pattern for a cantilever beam subjected to bending and rotation, and (b) experimental interferogram

field under study $w(x,y)$ on a known linear displacement field $w^*(x,y)$ [5.55]. The pattern corresponding to the $w^*(x,y)$ field can be produced by any of the methods described in Sect.4.2.1. In sandwich-holographic interferometry such a fringe pattern is obtained by properly rotating the sandwich holograms.

The concept underlying this method can best be illustrated in the example of bending of a cantilever beam. Here the interference fringes are orthogonal to the beam axis and, hence, $\partial w(x)/\partial y = 0$. Another fringe pattern $w^*(y)$ is produced in the direction of the y axis. Figure 5.18a shows schematically the fringe system created under these conditions, and Fig. 5.18b an experimental interferogram. Since the interference fringes are actually points of equal deflection for the total field, $w(x)+w^*(y)$, the derivative in the fringe direction is zero. If the fringe direction is specified by the angle α (Fig.5.18a), then the derivative of the deflection along the beam axis can be found from

$$\frac{\partial w(x)}{\partial x} = C\tan\alpha \qquad (5.34)$$

where $C = \partial w^*(y)/\partial y$ is determined by the given angle of rotation about the y axis. The value of C can be conveniently derived from the fringe pattern characterizing $w^*(y)$ on an undeformed part of the beam or on a special witness. As seen in (5.34), the error inherent in the derivative calculation depends on the fringe angle. The optimum value of the angle is close to 45°. By properly varying the rotation, one can adjust the angle to the optimum value in different parts along the beam.

The fringe pattern characterizing the total field $w(x)+w^*(y)$ can also be used to estimate the second derivative of deflection [5.56]. For this purpose one has to know the fringe curvature

$$\frac{\partial^2 w(x)}{\partial x^2} = \frac{\lambda}{2Rh\cos\alpha} \tag{5.35}$$

where h is the fringe spacing along the beam axis. Obviously, because of the large error involved in determining the radius R, (5.35) can only be used for a rough evaluation of the magnitude and the sign of the second derivative.

In a fringe pattern corresponding to $w(x,y)+w^*(x,y)$ one can relatively easily find the points where the gradient of the total field is zero. Then the derivatives of the original deflection field $w(x,y)$ can be unambiguously determined from the parameters of the fringe pattern due to the additional field $w^*(x,y)$,

$$\frac{\partial w(x,y)}{\partial x} = -\frac{\partial w^*(x,y)}{\partial x} , \quad \frac{\partial w(x,y)}{\partial y} = -\frac{\partial w^*(x,y)}{\partial y} . \tag{5.36}$$

In other words, the rotation of a local area on the object surface in the vicinity of the point of interest is compensated for by the additional displacement field. By properly adjusting the parameters of the fringe pattern characterizing the additional field (i.e., the direction and magnitude of rotation) one can compensate for the angle of rotation at any point on the object surface. The interferogram of the field $w(x,y)+w^*(x,y)$ is the finite-width fringe pattern.

An infinite-width fringe pattern obtained for the case of bending of a circular plate with a hole clamped along the edges and supported elastically at the center (Fig.5.6a) is shown in Fig.5.5. Figure 5.19 presents a finite-width fringe pattern obtained under deformation and rotation of the same plate. The four points *1*, *2*, *3*, and *4* are those positions where the equality (5.36) is met. The values of the derivatives are derived from the fringe pattern on the undeformed flange of the specimen.

The deflection derivatives in a given direction r can be determined from a fringe pattern obtained by loading the object and introducing a linear phase shift along r. The desired shift can conveniently be produced by displacing the photographic plates in the sandwich [5.57]. The finite-width fringe pattern obtained in this way is analogous to the one shown in Fig.5.19 but will be more crowded. The derivatives can be found from [5.57]

Fig.5.19. Finite-width interferogram of deflection of the plate presented in Fig.5.6a

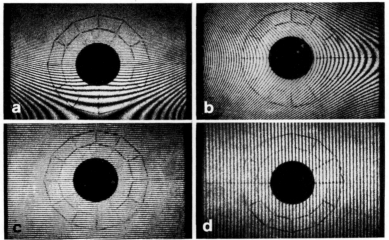

Fig.5.20. Interferograms of a plate with a hole subjected to bending (**a, b**); only linear phase shifts without plate loading (**c, d**)

$$\frac{\partial w}{\partial r} = \frac{\lambda}{2} \left[\frac{1}{\Delta r} - \frac{1}{\Delta r_0} \right], \qquad (5.37)$$

where Δr is the fringe spacing, and Δr_0 is the fringe spacing with a linear phase shift only.

This method is used for the determination of stresses along the edges of a hole in clamped plates loaded by a bending moment. A plate of D16T duralumin alloy measuring $6 \times 120 \times 300$ mm^3 has a 30 mm diameter central hole. An analysis of this problem (Sect.2.2.4) revealed the possibility of obtaining holographic interferograms at sufficiently high loading levels. A superposition onto the fringe pattern (Fig.2.24b) of a linear phase shift along the longitudinal axis of the plate yields the interferogram presented in Fig.5.20a, while a linear phase shift in the transverse direction results in the fringe pattern of Fig.5.20b. The interferograms corresponding to only linear

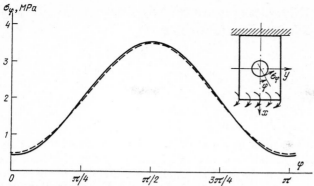

Fig.5.21. Distribution of stresses along the hole contour in a plate subjected to bending

phase shifts without plate loading are depicted for the same two cases in Fig.5.20c and d.

By using (5.37), the interferograms of Figs.5.20a and c can be used to determine the deflection derivatives along the edges of the hole in the longitudinal direction ($\partial w/\partial x$), and the interferograms shown in Figs.5.20b and d, those in the transverse direction ($\partial w/\partial y$). The distribution of the derivatives along the edge thus found allows derivation of the deflection derivative with respect to radius ($\partial w/\partial r$) and angle ($\partial w/\partial \varphi$). The distribution of the stress along the edge, σ_φ, can be determined by means of (5.16) by carrying out numerical differentiation of the ($\partial w/\partial \sigma_\varphi$) relation. The values of σ_φ deduced for a unit load are exhibited in Fig.5.21 with a solid line. The dashed line depicts the corresponding analytic relation.

6. Displacement Measurements on Objects under Elasto-Plastic Deformation

The appearance of residual deformations in machine parts and structural elements may dramatically affect their mechanical behavior and, hence, the functioning of a machine as a whole. Therefore, the possibility of revealing irreversible, residual changes on the surface of an object is of considerable practical importance. Information concerning the conditions conducive to the onset of residual deformation on the surface of a machine part can also be significant in the analysis and refinements of the calculational techniques employed.

The application of holographic interferometry opens up new possibilities for both purely qualitative and quantitative analysis of the residual strain fields on the surfaces of real objects. Here, we consider the holographic methods of measuring residual displacements, which allow establishment of the onset of residual deformation with high precision, following the kinetics of its development, and resolving the individual components (both elastic and plastic) under elasto-plastic deformation.

The possibility of measuring residual deformation with part of the object removed allows calculation of residual stresses in a machine part, which may have been created in the course of its manufacture or in plastic deformation. Residual stresses substantially affect the operation and reliability of a structure. The present chapter will focus on their determination and control, a major problem in solid mechanics. The use of holographic interferometry for studying residual deformation of microelements on the surface of an object when acted upon by a mechanical contact, cavitation-induced erosion, corrosion, or friction by measuring the changes in fringe contrast are also discussed. We believe that the examples chosen for illustration demonstrate graphically specific features of the method of solution of the above-mentioned problems.

6.1 Measurement of Residual Strain

6.1.1 Measurement of Residual Displacement by Holographic Interferometry

The holographic methods of measuring residual displacements are based on the interference of the light waves scattered by the surface of an object in its initial state, after loading and unloading. However, the practical implementation of these methods has additional difficulties compared to the techniques of measuring displacements in the elastic domain considered earlier. In esssence, the effect of rigid displacements of the object as a whole must be excluded in the situation where residual deformation appears only after the application of loads considerably larger than those required to study the elastic deformation of the object. In principle, these difficulties can be overcome in the same way as in Sect.2.2.4.

Another problem is that the range of measurable displacements is limited by the changes in the surface microstructure in the course of plastic deformation. Indeed, changes in the surface microrelief result in microsctructures on the object, which become decorrelated in the initial and final state, thus degrading the fringe contrast to the point of complete disappearance. Obviously, this phenomenon will manifest itself most strongly in areas of well developed plastic deformations and regions of their highest concentration, for instance, near crack tips. However, in the stage of initial plastic deformation, changes in the surface microrelief will not, for most structural materials, be so dramatic as to affect the holographic fringe contrast. Of course, studying the process of plastic, as opposed to elastic, deformation prevents repeating an experiment on the same sample under the same conditions.

Residual displacements of the object surface can be studied by real-time, two-exposure, or sandwich holography. In the real-time method, the hologram of the initial state of a region or of the whole surface is recorded. Then the object is loaded stepwise and subsequently unloaded. As the load increases, the frequency of the fringes observed through the hologram on the object surface also increases with the fringes finally merging and becoming unresolvable at sufficiently high loads. If the object is still in the elastic region, no fringes are observed after unloading. Their appearance at a certain loading-unloading stage is evidence for irreversible residual displacements. By analyzing these fringes and the conditions of their localization, one can conclude whether they have originated from rigid displacements of the object or from plastic deformation. In this analysis, one should pay particular attention to the conditions under which the translation- and rotation-induced fringes form. The presence of a rigid displacement is indicated by the appearance of fringes in the areas of the object surface where one cannot expect residual strains to occur either under the given loading conditions or the surface geometry.

In real-time holography one can select loading increments to provide an optimal fringe frequency for measuring residual displacements. As the loading increases and the limiting resolvable-fringe frequency is reached, a hologram corresponding to the new state of the object surface is recorded, after which the stepped loading and unloading is repeated.

An example of the application of the real-time method to establish the onset of plastic deformation in a body, was given in [6.1] by *Wernicke* and *Frankowski*. They determined critical contact loads of two steel discs leading to residual strains under a carburized case of different thickness. A special kinematic fixture ensured precise repositioning of the hologram in the optical setup, as well as its rotation and translation to compensate for the spurious fringes caused by rigid displacements of the discs under loading.

In the two-exposure method, the initial state of the object surface is recorded during the first exposure, and in the second exposure, the surface state after the object's loading and unloading. By recording two-exposure holograms with successive increases of the load, one can establish the moment of appearance of the fringes associated with residual strain and, hence, the point of transition of the object from the elastic to the elasto-plastic deformation domain. While measuring residual strains by the two-exposure method is more time consuming than that by the real-time technique, two-exposure holograms, as previously pointed out, are more convenient in quantitative analysis of interferograms, particularly when it is desired to determine two or three displacement-vector components. One should bear in mind that in the absolute fringe-order readout, the position of the zero-motion fringe can be reliably derived either by means of elastic strings or from the regions of the object where no residual displacements can take place.

Any two-exposure experiment should be preceded by an evaluation of the critical load at which plastic deformation is expected to begin, and by selecting the load increments which ensure sufficient fringe frequency and good fringe resolution. One should also bear in mind that the process of the building of residual strains is nonlinear in the general case and, hence, the load should be incremented in steps depending on its level. By comparing all possible states of the object surface at the various stages of deformation, one can select interferograms with an optimum fringe frequency. This can be done by using a series of masks which expose only certain areas of the photographic plate in a desired sequence [6.2]. Although this makes the experiment more complicated, the respective technique eliminates loss of information involved in measuring residual displacements.

The two-exposure method, just as the real-time approach, permits the process of residual deformation in structural elements to be followed in time. Two-exposure holography has been used to study the rolling conditions for the roller bands of a continuous casting machine for different degrees of their tightness [6.3]. The studies were carried out on models made on a 1:3 scale of 34XH1M steel, which is also used to manufacture the full-scale bands. The loading of the model is illustrated by Fig.6.1. Roller *1* is fixed rigidly on one side in the baseplate of the loading fixture mounted on the holographic bench; the other end rests through a hinge on support *2*. The load P is applied to band *3* through a rigid plane die *4*. The

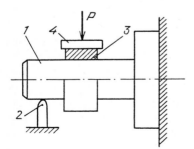

Fig.6.1. Loading of a roller band

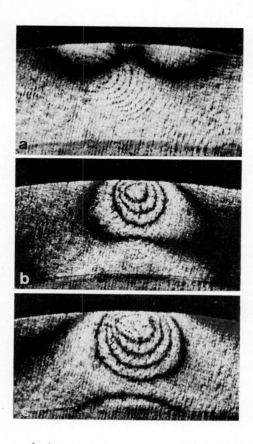

Fig.6.2. Interference fringes on the band end face in the (a) initial, (b) intermediate, and (c) final stages of loading

optical arrangement presented in Fig.4.4 was used to record focussed-image holograms of the end face of the band near the point of load application. The selection of the optical arrangement improving the fringe resolution was predetermined by the high degree of strain concentration in the region under study. As the loading increases, the residual deformation propagates away from the outer region of the band (Fig.6.2a) through its thickness (Fig.6.2b, c) and is considered critical in the last case, since at still higher loads the band decouples from the roller. The magnitude of the critical load is strongly affected by the tightness δ of the roller-to-band coupling whose optimum value was found to be $\delta/D = 0.0015$, D being the roller diameter.

In contrast to the above problem, the areas of the highest residual displacements frequently do not coincide with the regions where plastic deformation occurs. Indeed, the sites where a bent plate undergoes a maximum deflection may reveal only a slight plastic deformation or none at all. It is instructive to consider the development of residual strains in specific examples. *Yakovlev* et al. [6.4] studied the deformation of a flange cap loaded by an inner pressure q (Fig.6.3). The cap, made of D16T alloy, has a diameter D = 70 mm and h = 2 mm in the operating region, and two symmetrically located blind holes 22 mm in diameter at a distance a = 33 mm. The

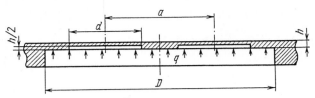

Fig.6.3. Loading of a flange cap

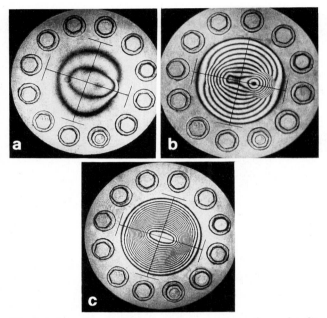

Fig.6.4. Interference fringe pattern on the surface of a flange cap revealing (a) first residual deflection, and development of (b) residual and (c) elastic deformation

displacements of points on the cap surface being normal to it, the optical arrangement shown in Fig.4.1 was chosen for the hologram recording.

Residual displacements are observed to appear only after the cap has been loaded by $q = 35$ MPa. The fringe pattern obtained under these conditions is displayed in Fig.6.4a where one sees two fringes corresponding to a maximum permanent deflection of 0.63 μm. On can follow the development of residual strains by analyzing the fringe patterns produced at higher loads. Figure 6.4b shows an interferogram obtained after a loading by the pressure $q = 45$ MPa. In contrast to the preceding pattern, the permanent deflections are seen to be distributed nonuniformly and localized primarily near one of the blind holes. To explore this further we analyze the character of elastic deformation of this cap as well. Figure 6.4c displays a fringe pattern obtained with the pressure increased from 10.0 to 10.5 MPa; it is seen to differ substantially from the previous patterns. The distributions of the elastic and permanent deflections within the operating area of the cap along

162

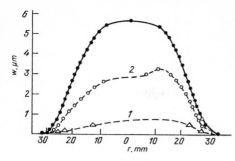

Fig.6.5. Elastic (solid line) and residual (dashed line) deflection curves on the working area of the flange cap after applying a pressure q: (*1*) 35 MPa, (*2*) 45 MPa

an axis passing through the blind hole centers are shown in Fig.6.5 and are different for the elastic and initial permanent deflections.

The possibility offered by holographic interferometry to visualize the residual displacement fields permits evaluation of the resistance of a material to initial plastic deformation which determines, in particular, the dimensional stability of structural elements [6.5]. Small strains should be measured on specimens subjected to bending under conditions ensuring high sensitivity and sufficient temperature stability. Uniform strain fields can be conveniently obtained on plane cantilevers in the form of either variable-width beams of equal strength loaded by a concentrated force or constant-cross-section beams loaded by a bending moment. Since the accumulated inelastic strain ϵ_x^r is essentially small, 10^{-4}, it can be determined in a surface layer via (5.11)

$$\epsilon_x{}^r = \frac{h}{2} \frac{\partial^2 w^r}{\partial x^2}, \tag{6.1}$$

where w^r is the deflection function along the specimen's axis x, and h is the specimen's thickness. The values of the stresses in the surface layers can likewise be found with good accuracy from the relations derived for elastic bending.

To exclude the effect of rigid displacements of a specimen, two-exposure holograms are obtained in an opposed-beam arrangement with the photographic plate holder mounted on the beam clamping fixture. Because the illumination and viewing directions are close to the normal to the specimen surface, one can use (4.1) for the determination of permanent deflections. By successively increasing the load and measuring the residual strains in each loading-unloading cycle, one can construct a residual deformation diagram (σ_x vs. ϵ_x^r relationship).

Plastic strain in extreme fibers, ϵ_x^p, can be determined using the expression [6.6]

$$\epsilon_x^p = \frac{1}{3} \frac{d\epsilon_x^r}{d\sigma_x} \sigma_x + \frac{2}{3} \epsilon_x^r. \tag{6.2}$$

163

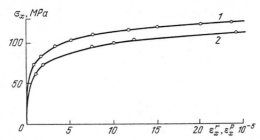

Fig.6.6. (*1*) Initial residual, ϵ_x^r, and (*2*) plastic, ϵ_x^p, deformation of M1 copper

Taking account of the fact that the residual deformation diagrams, when plotted on a logarithmic scale, represent piecewise-linear functions

$$\ln\epsilon_x^r = C_0 \ln\sigma_x + C ,\qquad (6.3)$$

where C_0 and C are constants, (6.2) can be rewritten in the convenient form:

$$\ln\epsilon_x^p = \ln\epsilon_x^r + \ln\frac{C_0+2}{3} .\qquad (6.4)$$

Initial residual and plastic deformation diagrams for Ml copper are shown in Fig.6.6

A method for measuring elasto-plastic deformation of plane bodies with reflection holography which uses the overlay interferometer discussed in Sect.3.2.4 is of special importance. It should be recalled that such an interferometer is particularly useful in determining in-plane displacement components. The application of an overlay interferometer permitted *Zhilkin* and *Gerasimov* [6.7] to study the elasto-plastic deformation of a 60×6 mm² plate with a 6 mm dia. central hole subjected to alternating loading. To avoid buckling under loading, the length of the specimen's operating region was kept within 70-80 mm. Two-exposure holograms observed at an angle Ψ = 18° (Fig.2.25) revealed speciment deformation at each loading step. Twenty-three interferograms were recorded in one cycle. Figures 6.7a,b display the fringe patterns for the first and eighteenth steps of the first loading cycle, respectively. Figure 6.8 presents, for the same cycle, a load dependence of the deformation in the axial direction at the most stressed point on the edge of the hole shown in Fig.5.3. The numbers specify the loading stages. This example suggests that holographic interferometry may be used in evaluating the resistance of a material to low-cycle loading. The application of low-modulus materials permits the study of complex metal-working processes. Thus, for example, models made of plastics or lead were employed to investigate elastic and plastic deformation in rolling [6.8].

Combining two-exposure and real-time methods increases the efficiency of holographic interferometry in the measurement of residual displacements. Here the initial state of the object under investigation is re-

Fig.6.7. Fringe patterns for (a) the first and (b) eighteenth steps of the first loading cycle

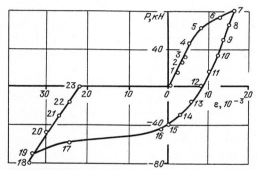

Fig.6.8. Deformation vs. loading curve for the first loading cycle

corded on two photographic plates. One of them is repositioned after processing into the optical setup and is used to study, in real time, the change of the fringe pattern under different loads with subsequent unloading. After a pattern of desired frequency has been obtained, the second photographic plate is exposed for the second time. The two-exposure hologram thus produced is subsequently utilized for the quantitative analysis.

Holographic methods for residual displacement measurement allow evaluation of the character of deformation of structural elements and establishment of the corresponding critical loading conditions for small displacements. Such studies may be viewed as a version of nondestructive testing.

6.1.2 Determination of Elasto-Plastic Strain Components

The two-exposure method allows separate measurements of the fields of the total and residual displacements of an object's surface. The required pairwise comparison of two light waves can be conveniently carried out by considering two regions on the same photographic plate [6.9, 10]. For this purpose one should select a holographic setup in which the light scattered by each point of the surface under investigation is uniformly distributed over the entire hologram. The exposure of a photographic plate conven-

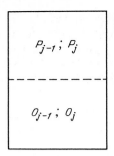

Fig.6.9. Exposure of photographic plate when recording elasto-plastic and residual deformation increments

$$P_{j-1} ; P_j$$

$$0_{j-1} ; 0_j$$

tionally divided into a lower and an upper part is shown schematically in Fig.6.9 for a jth loading step. The light wave from the unloaded object 0_{j-1} is recorded in the first exposure in the lower part of the photographic plate. In the second and third exposures, the upper part of the plate records the waves scattered by the object's surface at loads P_{j-1} and P_j, respectively. The wave 0_j, scattered by the object after full unloading, is recorded in the lower part of the photographic plate in the fourth exposure. Thus, two parts of the same photographic plate can be used to record two doubly-exposed holograms.

The fringe pattern observed in the reconstruction of the waves recorded on the upper hologram contains, in the general case, information on the increment of the total strain with the load changed from P_{j-1} to P_j, whereas the one reconstructed by the lower two-exposure hologram characterizes residual deformation. The absence of any fringes in the reconstruction made with the lower hologram suggests that at the given loading stage only elastic deformation occurs. The loading at which one begins to observe fringes through the lower hologram corresponds to a transition from the elastic to the elasto-plastic deformation. If the loading is incremented in equal steps, then it is also possible to compare the fringe patterns obtained at different loading levels. This allows examining the magnitude and the process of development of elastic deformation, the boundary of the elasto-plastic domain, as well as the character of development of residual deformation. By combining this information one can form a reasonable idea of the kinetics of deformation for the object under study.

A quantitative treatment of interferograms must take into account the change in the viewing direction for the upper and lower parts of a hologram. The modulus of the displacement vector **d** at a point on the object surface taken for a given load is obtained by summation of the corresponding displacement increments occurring at individual loading steps:

$$\mathbf{d} = \sum_{j-1}^{m} \mathbf{d}_j \qquad (6.5)$$

where $\mathbf{d}_j$ is the displacement increment at the jth loading step, and m is the number of loading steps.

If m is greater than the number of the loading steps m_r at which residual deformation occurs, then (6.5) can be used to calculate the displacement under elasto-plastic deformation, while for $m < m_r$ one determines the elastic displacements $\mathbf{d}^e$ only. The residual displacements $\mathbf{d}^r$ are obtained by summing up the increments of the residual displacements starting with the $m_r{}^{th}$ step

$$\mathbf{d}^r = \sum_{j=m_r}^{m} \mathbf{d}_j{}^r , \qquad\qquad (6.6)$$

where $\mathbf{d}_j{}^r$ is the increment of residual displacement at a loading step ($j \geq m_r$). The number of loading steps should be as small as possible since the measurement error increases with the number of steps used.

In the case of a large elastic strain, the number of the holograms can be reduced by recording a multi-exposure hologram in the upper part of the photographic plate (Sect.2.1.4). In this case blurred fringes show the transition to either nonlinear elastic or elasto-plastic deformation. This technique permitted *Shchepinov* et al. [6.11] to investigate the character of elasto-plastic deformation of gear teeth in the course of sizing, and to determine the parameters of the gear cutting tool profile. The sizing of a gear, or rather, its final grinding, is used for the final finishing of gear teeth and is an example of a technological process of cold deformation. After gear milling with some allowance along the tooth profile, the blank is rolled under the load in a mesh with a high precision tool gear. At the points where the blank is caught, the metal undergoes plastic deformation so that the tooth conforms to the required profile with the desired precision, removing the allowance. The most complex and cumbersome problem in this method of gear sizing is to find the proper profile of the tool gear or the distribution of the allowance over a blank's tooth. The problem becomes aggravated by the fact that the blank teeth undergo not only local plastic deformation, but bending as well. As a result, the compression of a gear's blank tooth is nonuniform in height, leading to a distortion of its profile.

The deformation of gear blank teeth was studied on commercial gears with a modulus of four and 36 teeth made of 45 steel. The tooth width was 28 mm with a constant allowance of 0.1 mm over the profile. The optical setup, displayed in Fig.6.10, permitted simultaneous recording of the side and end-face surfaces of a tooth by means of two holograms H_1 and H_2, respectively. To ensure appropriate illumination and viewing conditions for the side surface, two teeth were removed on each gear. The deformation was recorded by means of masks in the setup shown in Fig.6.9. Thus, four doubly-exposed holograms were simultaneously obtained at each loading step.

The fringe patterns observed in the reconstruction from H_1 reveals the number of contact points and the character of local deformation in the vicinity of these points, while the interferograms viewed through H_2 characterize the tooth bending. Residual deformations were determined on disengaged gears. The maximum number of the loading steps was less than ten.

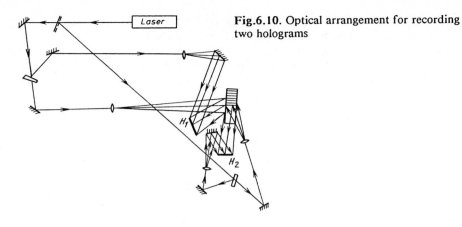

Fig.6.10. Optical arrangement for recording two holograms

A typical series of fringe patterns obtained for the first three loading steps, for one of the fixed gear conditions, is depicted in Fig.6.11. The blank and tool-gear axes were in the same plane with a skew of 0.15±0.05 mm per tooth length. Such a skew simulates the deviation in the tooth direction likely to occur in the gear milling process. The four columns of photographs are denoted *1*, *2*, *3* and *4*. The interferograms in columns *1* and *2* were taken for the end-face plane of the tooth and characterize the increments of the total and residual strain, respectively. The blank-tooth photographs in the second column reveal fringes only in the vicinity of the contact points, which confirms the absence of any displacement of the blank as a whole in the course of loading. As seen from the interferograms of the tooth side surface, shown in columns *3* and *4*, and characterizing, respectively, the increments of the total and permanent deflections, an elastic deformation is accompanied in the third loading stage by the appearance of a permanent deflection. Judging from the character of the fringes in these patterns, the deflection is nonuniform over the tooth length. This is caused by misalignment between the blank and tool axes. Experiments performed on differently engaged gears permitted determination of the best way to redistribute the allowance over the blank tooth and to adjust the tooth profile of the tool gear. The improved allowance decreased toward the tooth tip and the new tooth profile allowed engagement closer to the tip.

In addition to the scheme shown in Fig.6.9 which uses two regions of one photographic plate, one can also employ optical arrangements with one four-exposure hologram and two reference beams. In this case different states of the object surface, corresponding to the loads P_{j-1} and P_j, will be recorded with one reference beam, and two states, at the loads 0_{j-1} and 0_j, with the other. In such an arrangement both fringe patterns are observed from the same direction.

By recording elasto-plastic deformation at one loading step in three regions of a photographic plate it is possible to derive even more information than when only two regions are used [6.12]. In Fig.6.12 the letters *A*, *B*, *C* specify the three regions on the photographic plate, and the numbers *1*, *2*, *3* are the numbers of the exposures involved.

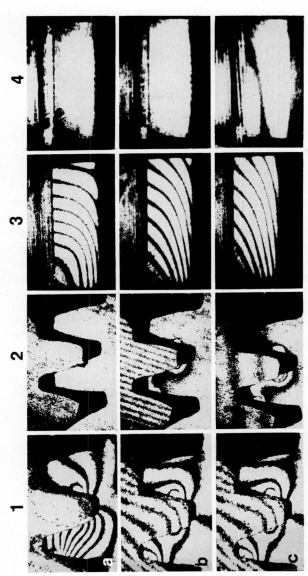

Fig.6.11. Interferograms of the end face (columns 1 and 2) and side (columns 3 and 4) surface of the gear teeth. (a) initial load $P_o = 0$, final load $P_f = 7.45$ kN; (b) $P_o = 7.45$ kN, $P_f = 13.35$ kN; (c) $P_o = 13.35$ kN, $P_f = 19.60$ kN

169

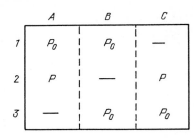

	A	B	C
1	P_0	P_0	—
2	P	—	P
3	—	P_0	P_0

Fig.6.12. Recording the components of elasto-plastic deformation

In the first exposure, the surface state at the initial load P_0 is recorded in the regions A and B. The state corresponding to the final load P is recorded in the regions A and C during the second exposure. The third exposure is made in the regions B and C after unloading the object back down to the initial state P_0. Thus, on one photographic plate using three exposures, three doubly-exposed holograms are recorded, which can be used to compare different states of the object surface under study. Reconstruction of the waves recorded on the holograms A, B and C yields fringe patterns describing, respectively, the increment of the total $(P_0;P)$, residual $(P_0;P_0)$, and elastic $(P;P_0)$ strains.

The technique for separate recording of displacements of different type can be illustrated by the case of bending of a $50\times50\times0.6$ mm^3 square plate made of D16T alloy. It is clamped along the outer edges and along the edges of a 20 mm diameter central hole. The plate was subjected to a uniformly distributed load. The optical arrangement depicted in Fig.4.1 was used in recording the holograms. The fringe patterns obtained at loads of P_0 = 144.0 MPa and P = 145.6 MPa are presented in Figs.6.13a, b, c corre-

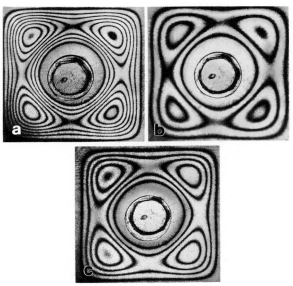

Fig.6.13. Interference fringes for the incrementation of (a) total, (b) residual, and (c) elastic deformation

170

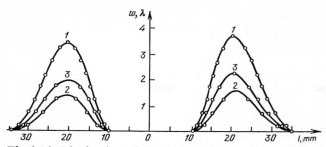

Fig.6.14. Distribution of (*1*) total, (*2*) residual, and (*3*) elastic deflections along the diagonal of a plate

sponding to the regions *A*, *B*, *C* on the photographic plate. One can readily see that the sum of the fringe orders at the point of maximum deflection for the elastic (n = 5) and residual (n = 3) displacements is equal to the fringe order in the total displacement interferogram (n = 8).

Deflection curves derived from these interferograms for one of the plate's diagonals are presented in Fig.6.14. The origin is at the plate's center. The total deflection W (curve *1*) and the sums of its components W^e and W^r (curves *2* and *3*) remain equal at all points. Some asymmetry exists in the two branches of the graph, which reflects the real conditions of deformation.

By using three reference waves, one can also separately record the displacement fields in the same single region of a photographic plate. In general, by properly choosing the number of regions to be used on a photographic plate and the number of exposures, one can study the deformation of an object over a broad range of loads with various relative contributions of the elastic and residual strain components. One such arrangement, also employing multi-exposure holographic interferometry, was used in testing the strength of a cylindrical pyrographite shell. An inner pressure was applied until the shell's fracture [6.13].

6.1.3 Recording of the Temporal Deformation Process

Comparison of fringe patterns obtained at different moments in time permits one to record the temporal development of a deformation. Usually the real-time and two-exposure methods of holographic interferometry are used for this purpose.

When the time intervals involved are on the order of hundreds of hours, particular attention should be paid to maintaining constant temperature and humidity of the environment and stable laser operation. One must make sure that the interferometer components remain in fixed positions with respect to one another. Otherwise, the fringe pattern obtained will reflect not only the deformation of the specimen, but also the variation of the parameters. To improve the stability of the interferometer setup, one must reduce the number of components to a minimum and fasten them as reliably as possible.

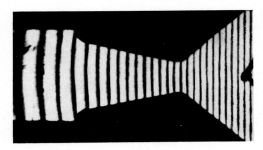

Fig.6.15. Holographed microcreep of a copper specimen in 80 hours

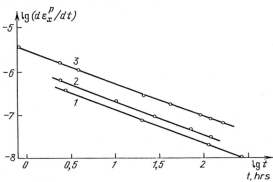

Fig.6.16. Copper creep rate diagram for σ_x (*1*) 65, (*2*) 85, and (*3*) 120 MPa

By successively recording doubly-exposed holograms with different time intervals between the exposures, one can construct creep curves of a material at the initial stage of small deformations (microcreep). In such experiments one should make sure that the duration of each exposure is substantially less than the period in which the deformation is recorded.

Creep deformation can be studied in the opposed-beam arrangement which has the smallest possible number of components. In testing a cantilever beam of Ml copper loaded by a constant concentrated force, the exposure time for a PE-2 (PFG-03) photographic plate was 10 s, and the intervals between exposures were varied from 30 min to 180 h. If the time between the exposures exceeded 10 h, the photographic plate mounted in the kinematic fixture (Sect.2.1.3) was removed from the optical setup after the first exposure, and the laser was switched off. After a predetermined period of time sufficient for the laser to reach the required operating condition, the photographic plate was repositioned in the setup, and a second exposure was made. A typical interferogram for a time interval of 80 h is presented in Fig.6.15. Figure 6.16 displays creep-rate curves for three values of stress obtained at the characteristic points of the copper microcreep diagram (Fig.6.6) plotted on a logarithmic scale. The plastic strains were determined, similarly to (6.1), by

$$\epsilon_x^p = \frac{h}{2} \frac{d^2 w^p}{dx^2} .$$

The linear behavior of the graphs suggests a power-law dependence of the creep rate with time.

When the time intervals involved are short (less than 10h), two-exposure recording of the deformation process can be made on separate regions of the same hologram [6.14]. In opposed-beam configurations, during the first exposure, one records the whole surface of the specimen in the initial state, and during the second exposure, only a part of the surface, visible through a slit mask. By moving the mask in the direction perpendicular to that of the specimen's tension, one can obtain a series of interferograms corresponding to the deformed states at different times. This technique improves the accuracy of determination of plastic strain since at each stage, the net strain is measured. The measurable strain is limited, however, by the fringe resolution in the last interferogram. This method was used to obtain creep curves on specimens of AG-4c unidirectional glass-fiber reinforced plastic [6.14].

Another way to study creep deformation is to successively expose different regions on a photographic plate by moving a mask. This is shown in Fig.6.17 where regions are denoted A, B, and C. The first exposure is made at time t_0 over the region A. At a predetermined time t_1 the second exposure is taken over the regions A and B. Subsequently, one records, in the regions B and C, the state of the specimen's surface at time t_2, and so on. Thus, the second exposure for the preceding hologram is performed simultaneously with the first exposure for the next one. As a result, one obtains m doubly-exposed holograms corresponding to m time intervals in m+1 exposures.

The same idea of successively recording two-exposure holograms of objects with a plane surface at different times can be also implemented with a focussed image configuration (Fig.6.18). A light wave scattered diffusely from the object surface 1 impinges on the prism 2 with mirror faces making an angle of 90°. The objective lenses 3 are mounted in the planes of the two holograms H_1 and H_2. Each of the photographed plates is illuminated by a plane reference wave. In this arrangement, the different areas on the photographic plate in Fig.6.18 are replaced by separate plates. The procedure is performed in the following order. At time t_0 the photographic plate H_1 is exposed, followed by the exposure of H_1 and H_2 at time t_1, after which the plate H_1 is replaced by a new plate H_3, exposed at time t_2, and so on. Because it is based on the use of masks, just as the preceding method, this scheme prevents loss of information about the creep which may occur when a photographic plate is replaced.

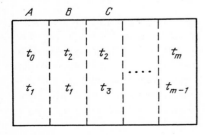

Fig.6.17. Continuous recording of two-exposure holograms on one photographic plate

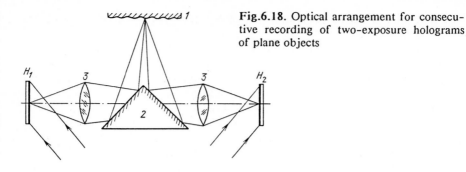

Fig.6.18. Optical arrangement for consecutive recording of two-exposure holograms of plane objects

Combined application of the real-time and double-exposure techniques increases the amount of obtainable information relative to that given by each technique separately. For example, in a study of the creep in localized areas of a cylindrical shell made of a composite material, with real-time holography one could measure the displacements and with the double-exposure technique, the creep rate [6.15].

Holographic interferometry can also be used in studies of processes other than creep. *Tsilosani* et al. [6.16] investigated the development of shrinkage deformation in drying concrete; *Aleksandrovskii* and *Shtan'ko* [6.17] have examined low temperature-induced fracture of cellular concrete. In these studies they succeeded in recording the formation and propagation of cracks.

6.2 Determination of Residual Stresses

6.2.1 Surface-Etching Method

Layer-by-layer removal of material permits one to determine uniform residual stresses in each layer of a specimen [6.18]. To facilitate both the experiment and the subsequent stress calculation, one usually takes specimens of a simple shape, namely, cantilever beams or plates of rectangular cross section. In cases where this is possible, the material should be removed by etching, which can be done either directly in the holographic setup [6.19] or outside it, using a kinematic fixture to return the specimen back into the initial position. After the removal of a layer, the residual displacements of the object surface are measured by either the real-time or the double-exposure technique. Obviously, the surface to be holographed is opposite to the one subjected to etching.

To calculate residual stresses, we consider the condition of an elastic equilibrium of the part of the object left after etching of a layer [6.18]. The destruction of a part of the object by etching is assumed to be equivalent to the application of residual stresses to the remaining part. The stresses which acted on the etched-off layer and those in the remaining part are of opposite sign. Provided that the stresses do not exceed the elastic limit, from the experimentally measured displacements or strains of the remaining part,

174

one can deduce the residual stresses in the etched-off layer by conventional theory of elasticity.

For a cantilever beam of length ℓ and a constant cross section of height H, the normal residual stresses σ_x^r at a level h_s from the surface can be calculated from the measured maximum deflections f(h) caused by successive removal of layers dh down to this level [6.18]

$$\sigma_x^r = \frac{E}{2\ell^2}\left[(H-h_s)^2 \frac{df(h_s)}{dh} - 4(H-h_s)f(h_s) + 2\int_0^{h_s} f(h)dh \right] ,$$

where E is the elastic modulus of the material, and $df(h_s)/dh$ is the derivative of the maximum deflection function for a given level h_s. In cases where $H/h_S > 50$, the stress calculations can be limited to the first term of the expression in brackets. Bearing this in mind, and removing only one layer of thickness h, the above expression reduces to

$$\sigma_x^r = \left(\frac{E}{3\ell^2}\right) \frac{H^2 f}{h} \qquad (6.7)$$

where f is the maximum deflection of the specimen. Equation (6.7) can be rewritten in terms of the beam curvature χ:

$$\sigma_x^r = (E/6) \frac{H^2 \chi}{h} . \qquad (6.8)$$

This method for determining the mean stress in the surface layer of a specimen can be used to find residual stresses in films [6.20]. Figure 6.19 presents a two-exposure fringe pattern typical of a cantilever beam obtained by etching an h = 0.5 μm oxide layer off a silicon wafer of thickness H = 0.5 mm. The oxide layer was etched off by fluoric acid between the exposures. The fringe pattern obtained here gives an idea of the character of the deformation of the specimen surface as a whole and, thus, permits judgement on the distribution of the residual stresses, in particular their uniformity.

Figure 6.20 shows the derivative of the residual deflection dw/dx along a specimen length, revealing a linear region of constant curvature. It

Fig.6.19. Reconstructed image of a substrate after etching

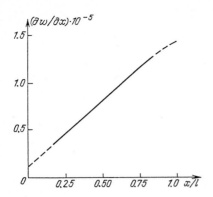

$(\partial w/\partial x)\cdot 10^{-5}$

Fig.6.20. Distribution of the derivative of residual deflection along a specimen

is within this region that the experimental data are in best agreement with the theoretical calculations for the stress determination. Thus, to improve accuracy and reliability of the determination of residual stresses one should obviously use information on the deformation of this part of the specimen only. The dimensions of this region are dependent on the actual boundary conditions and the level of residual stresses. The sign of the deflection which unambiguously determines the sign of the residual stresses is found by means of the method described in Sect.4.1.2.

This technique has been applied to evaluate the effect of various technological and operating factors on the residual stresses in films of systems used in microelectronics. For example, it has been found that neutron irradiation at a fluence of 10^{13} n/cm^2 reduces the tensile residual stresses in films of the Si/SiO_2 system by about 20% [6.21].

The expressions derived for the calculation of residual stresses from deflections of a specimen are also valid in the cases where thin films are deposited on it rather than etched away. This direct determination of residual stresses may be considered to be a method of nondestructive stress testing, originally demonstrated by *Ramprasad* and *Radha* [6.22] under such conditions to measure residual stresses in evaporated films.

Holographic interferometry also permits determination of residual stresses which vary over the specimen length [6.23, 24]. The out-of-plane $\sigma_x^r(x, h_s)$ and in-plane $\tau_{xy}^r(x, h_s)$ stresses in a cross section x can be found with the expressions

$$\sigma_x^r(x, h_s) = -\frac{E}{6}\left[(H-h_s)^2\frac{\partial \chi}{\partial h}(x, h_s) - 4(H-h_s)\chi(x, h_s) + 2\int_0^{h_s}\chi(x, h)dh\right],$$

$$\tau_{xy}^r(x, h_s) = -\frac{E}{6}\left[(H-h_s)^2\frac{\partial \chi}{\partial x}(x, h_s) - 2(H-h_s)\int_0^{h_s}\frac{\partial \chi}{\partial x}(x, h)dh\right],$$

where $\chi(x, h_s)$ is the beam curvature at the cross section x after removal of the layer with thickness h_s. The curvature and its derivative with respect to length are related to the deflections $w^r(x)$ through the relations

176

$$\chi(x, h_s) = \frac{\partial^2 w^r}{\partial x^2}, \quad \frac{\partial \chi}{\partial x} = \frac{\partial^3 w^r}{\partial x^3}.$$

Obviously, if the curvature is constant over the specimen length, then its derivative is zero, and no in-plane stresses can be present.

6.2.2 Hole-Drilling Method

The method involving drilling of holes [6.25] is currently widely used to determine residual stresses at given points of an object surface. According to the model underlying this method, drilling a hole in the region of interest causes deformation in the adjoining regions which are acted upon by stresses opposite to the residual ones. The strain data are used to calculate the stresses assumed to correspond to the hole's center and are taken to be uniform over the hole's area. Obviously, while the calculation error will decrease with decreasing hole diameter, the strains will also be reduced and, hence, a loss of accuracy in their measurement will result.

Holographic methods of strain measurement overcome these contradicting factors and are thus widely employed in determining residual stresses in various modifications of the hole-drilling technique [6.26-28]. This technique, to a varying degree, deals with the key problem of accuracy of the strain-induced displacement measurement such as location of the zero-motion fringe, accounting of rigid displacements, and isolation of the individual components of the displacement vector. A proposal to increase the measurement sensitivity threshold through the use of higher-order waves in a modification of the focussed-image holography has been put forward [6.29].

Depending on the actual problem to be solved, either blind or through holes can be used. Through holes are drilled in thin-walled structures, usually with a diameter not exceeding the specimen thickness. Blind holes are made in massive objects. A typical fringe pattern obtained after drilling of a through hole is shown in Fig.6.21 displaying the out-of-plane components of the displacement vector. Note the symmetry of deformation with respect to the two orthogonal axes coinciding with the principal stress axes. This coincidence is utilized in the following discussion of evaluating the stressed state.

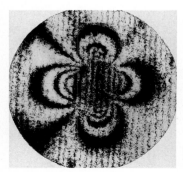

Fig.6.21. Interference fringe pattern around a through hole

The principal residual stresses σ_x^r and σ_y^r due to drilled through holes are related to the displacement w^r normal to the object surface by the expression

$$\sigma_x^r - \sigma_y^r = \frac{Ew^r}{\mu h \cos 2\theta} \qquad (6.9)$$

where E is the elastic modulus of the material employed, μ is Poisson's coefficient, h is the plate thickness, θ is the angle between the x axis and the radius connecting the point in question at the edge of the hole with its center, the x and y axes pass through the hole's center. Equation (6.9) refers only to the maximum in-plane stresses, so that in order to separate the principal stresses, one has to have additional information on the relationship between them. This can be obtained in some cases. For example, the stressed state of a long straight weld seam is close to uniaxial ($\sigma_y^r = 0$), and (6.9) can be employed to derive the residual stresses σ_x^r along the seam axis.

In the most general case for stress determination using blind holes, the calculation of residual stresses requires an analysis of the stressed state within the framework of the three-dimensional theory of elasticity. Its solution by the finite-element technique has resulted in the development of a convenient method to calculate residual stresses by studying the displacements measured by methods of holographic interferometry [6.30].

We use the coordinate system (x, y, z) with the origin on the object surface at the center of the blind hole. The x and y axes coincide with the directions of the principal stresses σ_x^r and σ_y^r. The dependences of the out-of-plane, w^r, and in-plane, u^r and v^r, components of the displacement vector along these coordinate axes on the unit residual stresses $\sigma_x^r = 1$ and $\sigma_y^r = 1$ were found by the finite-element technique for different ratios of the hole's depth to its radius. We introduce the following shorthands for the unit residual displacements at $\sigma_x^r = 1$, $\sigma_y^r = 0$:

$$w_{1x}^r, \quad w_{1y}^r, \quad u_1^r, \quad v_1^r,$$

and, similarly, for the case of $\sigma_y^r = 1$, $\sigma_x^r = 0$:

$$w_{2x}^r, \quad w_{2y}^r, \quad u_2^r, \quad v_2^r.$$

Obviously we have pairwise equality of displacements

$$w_{1x}^r = w_{2y}^r, \quad w_{1y}^r = w_{2x}^r, \quad u_1^r = v_2^r, \quad v_1^r = u_2^r.$$

Using the principle of independence of the action of different stresses one can transform from unit stresses to initial stresses, which corresponds to the displacements w_x^r, w_y^r, u^r, and v^r. Thus, for the measurement of out-of-plane components of the displacement vector only, one can write the following system of equations

$$w_{1x}^r \sigma_x^r + w_{2x}^r \sigma_y^r = w_x^r \,, \qquad\qquad (6.10)$$

$$w_{1y}^r \sigma_x^r + w_{2y}^r \sigma_y^r = w_y^r \,.$$

Solution yields the residual stresses σ_x^r and σ_y^r. In a similar way one can construct three more systems for the measurement of the displacement components w_x^r and v^r, w_y^r and u^r, u^r and v^r.

Residual stress calculation based on the values of the out-of-plane components only is of limited use at large hole depths when the values of w_x^r and w_y^r approach one another, however. The system of equations making use of them becomes poorly conditioned. There are other systems of equations which do not have such limitations.

Of particular interest are shallow holes for which the stress in a given principal direction is determined by the displacements measured along the coordinate axis in the same direction. In this case, the computational formulas and, hence, the technique, are greatly simplified:

$$\sigma_x^r = \frac{w_x^r}{w_{1x}^r}\,, \qquad \sigma_y^r = \frac{w_y^r}{w_{2y}^r}\,.$$

Special, portable holographic equipment is available for applying the hole-drilling method to studies of residual stresses in local areas of various structures [6.31, 32]. One of such instruments, called "Limon" [6.31], and developed at the Institute of Applied Mechanics, Academy of Sciences of the USSR, is exhibited in Fig.6.22. The interferometer has been arranged so that its overall dimensions are minimized, and a high vibrational stability is ensured. The instrument is based around a LG-79-1 laser and is capable of recording out-of-plane displacements in the vicinity of a drilled hole on PE-2 (PFG-03) photographic plates. The exposure time is in the order of 10 s. Note that the conditions of deformation around a hole and, hence, the precision with which the residual stresses can be determined, depend on factors such as the drilling speed, the applied load, and the tool's wear [6.33].

Fig.6.22. Holographic installation "Limon"

6.3 Mechanical Contacts

6.3.1 Study of the Contact Surface

The conditions at the point where machine parts come in contact are primarily governed by the shape and size of the surface over which they interact and by the applied load, as well as the surface roughness and the mechanical properties of the material. The interaction of parts under load is invariably accompanied by an irreversible change of the surface microrelief over the contact area. As already pointed out, a change in the surface microstructure between holographic exposures leads to a reduction of fringe contrast. This effect is employed to determine contact regions in elastic and elasto-plastic interaction [6.34, 35].

The double-exposure method is particularly convenient for studies of the contact surface since it provides the simplest way of obtaining high-contrast fringes. In a first exposure one records the light wave scattered by the surface of the object in the region of assumed contact. Then the object is brought in contact with the other object under the required load and subsequently unloaded. After removing the objects in contact, a second exposure is made. To produce a fringe pattern localized on the object's surface, prior to the second exposure, either the object is rotated about an axis lying in the plane of its surface, or the source of illumination is shifted. Reconstruction of the waves recorded on such a hologram produces a pattern of fringes on the object surface with degrading contrast until complete disappearance is observed in the area of contact. This permits visualization of the contact region and determination of its dimensions.

Machine parts can interact either in an elastic way so that no residual macrodeformations in the vicinity of the contact will be present, or elasto-plastically, in which case the fringes in the contact area will become bent. An interferogram typical of an elastic contact of a steel plate with a steel sphere is presented in Fig.6.23 for three loads. The plate had a hardness of 30 HRC, the sphere was 600 mm in diameter, and had a hardness of 49 HRC. The roughness of the two surfaces in contact was $R_a = 1.5$ μm. The break in the fringes indicates the boundary of the contact area. In a contact of a steel plate with a hardness of 12 HRC with the same sphere, the character of deformation is very different, as demonstrated by Figs.6.24a,b. As

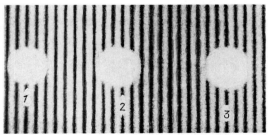

Fig.6.23. Breaks in the fringes in the reconstructed image of a plate after an elastic contact with a load P: (1) 39, (2) 49, and (3) 60 kN

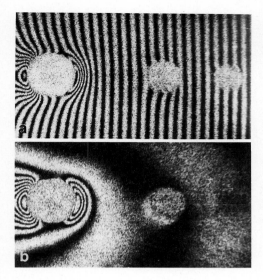

Fig.6.24. Reconstructed image of the plate's surface after an elasto-plastic contact in (a) finite and (b) infinite width fringes

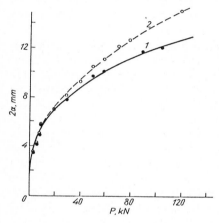

Fig.6.25. Diameter 2a of contact surface vs. load **P** for (*1*) elastic and (*2*) elasto-plastic interaction

the load increases, one sees the finite-width fringes in the vicinity of the contact become distorted and shifted, and new fringes appear in the infinite-width fringe interferogram. Figure 6.25 displays the experimentally obtained dependences of the contact-surface diameter 2a on load P for a sphere in contact with a plate (see above). For comparison, the solid curve shows the same relation obtained by solving Hertz's problem [6.36]. One readily sees good agreement between experiment and theory under the conditions of elastic contact, but a noticeable increase of the contact-spot diameter as one crosses over to the elasto-plastic domain.

This technique has been used to establish the conditions of contact interaction of a banded roller with a plate. The studies were carried out within the frame of the above mentioned problem [6.3] on the models shown in Fig.6.1. The fringe patterns appearing on the band surface are presented in Fig.6.26a,b for two loading steps. The breaks in the fringes,

Fig.6.26. Reconstructed image of the band surface for a distributed load of (a) 4.9 kN/cm and (b) 14.7 kN/cm

observed on the photographs, reveal a substantial nonuniformity of the contact area along the band's axis. The bending of the fringes near the boundary of the contact spot suggests that the interaction of the band with the plate is of elasto-plastic character.

Note that this technique permits visualization of the contact surface as a set of discrete points which correspond to the breaks in the fringes. Contact spots of small size are difficult to study in this way because of the limited number of reference fringes passing through them. This difficulty can be overcome by employing the holographic-image substraction technique based on the use of Fourier holograms [6.37].

Figure 6.27a illustrates the principle of recording Fourier holograms of an object. Recall that in this method one records the spatial spectrum of the object wave. Lens L_1 performs the Fourier transformation of the object field. Visualization of the contact surface is achieved in the following way. A first exposure of the surface area is made. Before the second exposure, the objects are brought in contact under load, and a phase shift is made by turning the reference wave through a small angle. As a result, the spatial spectrum of the image becomes modulated by equally spaced fringes. If we now illuminate the hologram with a nonexpanded laser beam in the region of a dark fringe (Fig.6.27b), then the region of the microrelief affected by

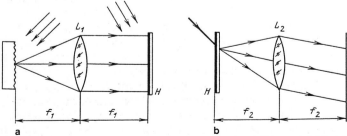

Fig.6.27. (a) Formation of Fourier hologram, and (b) spatial filtration of object field

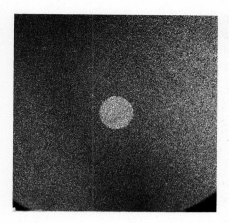

Fig.6.28. Visualization of a contact surface by holographic subtraction

contact interaction will be visualized against a dark background. In the arrangement shown in Fig.6.27b, lens L_2 makes the inverse Fourier transformation of the reconstructed field.

Figure 6.28 displays a typical fringe pattern observed in a contact of a steel ball with a plate. The boundary of the contact area is represented by a continuous line. The size of the spot obtained experimentally agrees with the value given by Hertz's expression [6.36].

6.3.2 Determination of Contact Pressure

If in a contact interaction of two objects with rough surfaces interference fringes are observed throughout the contact area, one can determine the distribution of contact pressures from the variation in the fringe contrast [6.38]. The dependence of fringe contrast on contact pressure was found by considering the problem of the elastic contact between the face end of an absolutely rigid cylinder and the plane surface of a plate. The solution is well known [6.39]:

$$q = \frac{P}{2\pi a\sqrt{a^2 - r^2}}, \tag{6.11}$$

where q is the contact pressure, P is the load, a is the cylinder radius, and r is the distance to the cylinder axis. As follows from (6.11), in the central region of the contact area, up to $r \simeq a/2$, the contact pressures are practically constant and equal to

$$q \simeq \frac{P}{2\pi a^2}. \tag{6.12}$$

An experimental study was made of the contact region of a thick plate with a hardness of 39 HRC with the end face of a cylinder of radius a = 15 mm and hardness 66 HRC. The roughness of the surfaces in contact was R_a = 1.5 μm. Figure 6.29 shows a fringe pattern on the plate surface with a load P = 300 kN obtained by the two-exposure technique with a variation of the angle of the illuminating beam. The photograph reveals a constant

183

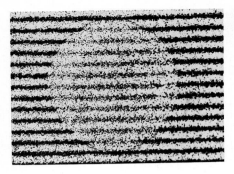

Fig.6.29. Interferogram of a plate in the area of contact with the end face of a cylinder

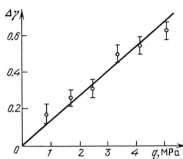

Fig.6.30. Fringe contrast change $\Delta\gamma$ vs. contact pressure q

fringe contrast over the central region which, however, is lower than that outside the contact area. At the boundary of the contact surface, the fringe contrast drops practically to zero, which qualitatively corresponds to infinitely high contact pressures when calculated by (6.11) for r→a.

The experimental dependence of the variation of fringe contrast in the central region of the contact area, $\Delta\gamma$, on contact pressure calculated by (6.12) is shown graphically in Fig.6.30. It is seen to be linear over a rather broad load range. If we assume that the linear character of the $\Delta\gamma$ vs. q dependence prevails over a broad range of roughness, hardness, and other characteristics of the contacting objects, then the distribution of pressures in the contact area can be also obtained without the preliminary construction of a calibration graph. Indeed, from the linear dependence we arrive at

$$q = C\Delta\gamma , \qquad (6.13)$$

while from the condition of equilibrium it follows that

$$P = \int_S q\,dS = C\int_S \Delta\gamma\,dS , \qquad (6.14)$$

where S is the contact surface area, P is the load, and the constant C can be derived by normalization. On finding C, one can determine the pressure distribution over the contact area. This method was employed to find contact pressures between a steel sphere with a radius of 500 mm and a steel

184

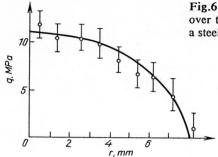

Fig.6.31. Radial distribution of contact pressure over the indentation produced by a steel sphere on a steel plate

plate. The hardness of the sphere was 44 HRC, and that of the plate 39 HRC. The roughness of the two surfaces was R_a = 1.5 μm. The chosen load, P = 140 kN, produced an indentation large enough to study, producing a clearly observable fringe pattern throughout its area. The distribution of the contact pressures along the radius of the indentation is presented in Fig.6.31. The solid curve shows the contact-pressure distribution obtained by solving Hertz's problem. One notices good agreement between the experimental data and the theoretical calculations, which yields indirect support to the validity of the linear relationship (6.13) under these conditions.

6.3.3 Study of Corrosion and Cavitation-Induced Erosion

The drop in fringe contrast resulting from irreversible changes in the shape of the elements in the surface microstructure can also be used to investigate other processes accompanied by decorrelation of the microstructure. This phenomenon was employed in corrosion studies [6.40, 41]. In their investigation of the corrosion of a brass plate exposed to nitric acid vapor, *Petrov* and *Presnyakov* [6.40] derived an analytic expression relating the variation of the contrast and fringe shift to the corrosion rate. The corrosion rate was determined experimentally from the variation of the fringe contrast only. Degradation of fringe contrast in a holographic interferogram has also been used to reveal defects in chips [6.42]. This technique was found to be capable of monitoring variation in the microrelief due to diverse causes such as corrosion, oxide film destruction, and mechanical damage. We have used this approach to study the surface erosion of hydraulic units induced by corrosion [6.43].

Consider the relation between the erosive action on the surface microstructure, and the contrast and shape of holographic fringes. We define A_1 and A_2 to be the complex amplitudes of the reconstructed waves before and after the erosive action

$$A_1 = a_1 \exp(-i\varphi_1) \ ,$$
$$A_2 = a_2 \exp(-i\varphi_2) \ , \tag{6.15}$$

where a_1 and a_2, φ_1 and φ_2 are the wave amplitudes and phases, respectively. For the intensity distribution of the fringe pattern we obtain

$$I = \langle |A_1 + A_2|^2 \rangle \qquad (6.16)$$

where the brackets denote averaging over the surface element marginally resolved by the given optical system. The size of this element substantially exceeds the transverse size of the micro-roughness (i.e., it is assumed that the optical system does not resolve the surface structure). We now assume that the phases φ_1 and φ_2 and amplitudes a_1 and a_2 are statistically independent, which gives, after some manipulation

$$I = \langle a_1{}^2 \rangle + \langle a_2{}^2 \rangle + 2\langle a_1 a_2 \rangle \langle \cos(\Delta\varphi) \rangle \qquad (6.17)$$

where $\Delta\varphi = \varphi_1 - \varphi_2$ is the phase shift caused by the variation of the surface microrelief originating from the erosive action.

We now consider the quantity $\langle \cos(\Delta\varphi) \rangle$. First we make the transformation from averaging over a specimen surface area with a large number of microstructural elements to that over an ensemble. To do this, we introduce a distribution function f(d), where d is the vector of the surface point displacement resulting from the erosive action. The vector d is related to the phase variation through the principal relation of holographic interferometry (2.41)

$$\Delta\varphi = \frac{2\pi}{\lambda}(\hat{e}_s - \hat{e}_1)\cdot d , \qquad (6.18)$$

where $\hat{e}_s$ and $\hat{e}_1$ are the unit illumination and viewing vectors for the surface point in question. Therefore,

$$\langle \cos(\Delta\varphi) \rangle = \int \cos\left[\frac{2\pi}{\lambda}(\hat{e}_s - \hat{e}_\perp)\cdot d \right] f(d) dd . \qquad (6.19)$$

Generally speaking, the function f(d) is unknown and Gaussian. This appears to be a valid assumption if we consider that in the course of erosive action, each point of the microrelief suffers a large number of displacements, i.e., that d is actually a sum of a large number of random terms. Then for f(d) we can write:

$$f(d) = f(d_\parallel, d_\perp) = \frac{1}{(2\pi)^{3/2} \Delta_\parallel{}^2 \Delta_\perp{}^2} \exp\left(-\frac{d_\parallel{}^2}{2\Delta_\parallel{}^2} - \frac{(d_\perp - \bar{d}_\perp)^2}{2\Delta_\perp{}^2} \right), \qquad (6.20)$$

where $d_\parallel$ and $d_\perp$ are the components of the vector d parallel and perpendicular to the specimen's surface, respectively. The quantity $\bar{d}_\perp$ is the displacement of the mean microrelief level due to erosion, $\Delta_\parallel$ and $\Delta_\perp$ represent the variances of $d_\parallel$ and $d_\perp - \bar{d}_\perp$, respectively. The quantities $\Delta_\parallel$ and $\Delta_\perp$ characterize the degree to which the surface microrelief becomes decorrelated by erosion.

By substituting (6.20) into (6.19) and integrating, we come to

$$\langle\cos(\Delta\varphi)\rangle = \cos\left[\frac{2\pi}{\lambda}(e_{s,\perp}-e_{1,\perp})\bar{d}_\perp\right]$$

$$\times \exp\left[-\frac{\pi\Delta_\perp^2}{\lambda}(e_{s,\perp}-e_{1,\perp})^2 - \frac{\pi\Delta_\|^2}{\lambda}(e_{s,\|}-e_{1,\|})^2\right], \quad (6.21)$$

where $e_{s,\perp}$ and $e_{1,\perp}$) are the out-of-plane, and $e_{s,\|}$ and $e_{1,\|}$, in-plane components of the $\hat{e}_s$ and $\hat{e}_1$ vectors, respectively. The first factor in (6.21) is associated with the shift of the central level of the microrelief and accounts for the variation of the fringe geometry in the interferogram. The second factor is due to decorrelation and describes the degradation of the fringe contrast [6.44]. The phase shift

$$\frac{2\pi}{\lambda}(e_{s,\perp} - e_{1,\perp})\bar{d}_\perp$$

is related to the variation of the fringe pattern through the expression (2.44)

$$\lambda n(e_{s,\perp} - e_{1,\perp})\bar{d}_\perp . \quad (6.22)$$

Introducing into (6.22) the angles of illumination α_s and observation α_1 which are measured from the normal to the object surface at the point in question, we find

$$\bar{d}_\perp = \frac{\lambda n}{\cos\alpha_s + \cos\alpha_1} . \quad (6.23)$$

Using (6.17,21), we obtain for the fringe contrast V

$$V = \frac{2(\langle a_1 a_2\rangle)}{\langle a_1^2\rangle + \langle a_2^2\rangle}\exp\left[-\frac{2\pi^2}{\lambda^2}(\cos\alpha_s+\cos\alpha_1)^2\Delta_\perp^2 - \frac{2\pi^2}{\lambda^2}(\sin\alpha_s-\sin\alpha_1)^2\Delta_\|^2\right].$$

$$(6.24)$$

We assume [6.40] that the erosion randomizes the phase but not the amplitude, so that $a_1 = a_2$. Then (6.24) can be written as

$$\ln\left(\frac{1}{V}\right) = \frac{2\pi^2}{\lambda^2}(\cos\alpha_s+\cos\alpha_1)^2\Delta_\perp^2 + \frac{2\pi^2}{\lambda^2}(\sin\alpha_s-\sin\alpha_1)^2\Delta_\|^2 . \quad (6.25)$$

Expression (6.25) contains two unknowns, $\Delta_\perp$ and $\Delta_\|$. They can be found by recording two holograms of the object at two different viewing angles, α_1 and α_2. Then one can write the following system of two equations with two unknowns:

$$\ln\left(\frac{1}{V_1}\right) = \frac{2\pi^2}{\lambda^2}[(\cos\alpha_s+\cos\alpha_1)^2\,\Delta_\perp{}^2 + (\sin\alpha_s-\sin\alpha_1)^2\,\Delta_\|{}^2]\,,$$

$$(6.26)$$

$$\ln\left(\frac{1}{V_2}\right) = \frac{2\pi^2}{\lambda^2}[(\cos\alpha_s+\cos\alpha_2)^2\,\Delta_\perp{}^2 + (\sin\alpha_s-\sin\alpha_2)^2\,\Delta_\|{}^2]\,,$$

where V_1 and V_2 denote the contrasts of fringes reconstructed from the hologram at the viewing angles α_1 and α_2, respectively. One can now solve (6.26) for $\Delta_\perp$ and $\Delta_\|$.

The above approach was checked experimentally on specially fabricated specimens which were subjected to erosive action by cavitation on a test bench of the water turbine laboratory at the Leningrad Metal-Working Plant. The stand was actually a flow-through apparatus, and the cavitation nozzle was a Venturi channel with a rectangular cross section. The angle at the exit from the slit measuring 4×60 mm^2 was 12°. Specimens 60×28 mm^2 in size, made of 1X18H10T stainless steel, were fixed to the side wall of the nozzle. The flow velocity in the nozzle was 39 m/s.

Doubly-exposed, finite-width fringe patterns of the specimens were recorded in a conventional two-beam arrangement. The specimens were mounted in a special holder permitting removal after the first exposure to subject them to cavitation-induced erosion with subsequent repositioning. Figure 6.32 shows a specimen mounted in this holder.

Finite-width fringe patterns of specimens tested on the cavitation stand measured at different times are displayed in Fig.6.33. They reveal that as the erosion time increases, the shape of the fringes changes and their contrast degrades. This shows both ever increasing deformation of the mean surface of the microrelief and its decorrelation.

The holograms depicted in Fig.6.33 were obtained with the optical setup displayed in Fig.4.1. Using the condition $\alpha_s = \alpha_1 = 0$, one can derive from (6.25) the value of $\Delta_\perp$:

$$\Delta_\perp = \frac{\lambda}{2\pi\sqrt{2}}\sqrt{\ln(1/V)}\,.$$

$$(6.27)$$

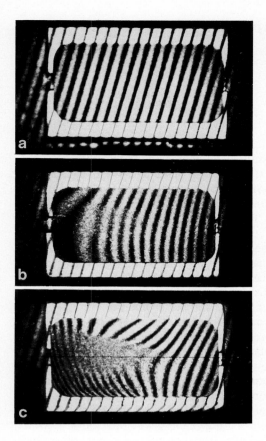

Fig.6.33. Reconstructed image of specimen surface (a) in its initial state, and after erosion for (b) 20 min, and (c) 40 min

The fringe contrast was determined from the reconstructed real images. Figure 6.34 shows a distribution of the quantity $\Delta_\perp$ along the line drawn in the pattern in Fig.6.33c.

Thus, holographic interferometry allows determination of the principal parameters characterizing the degree of surface erosion, namely, the distortion of the mean surface of the microrelief $d_\perp$ and the variances of the normal ($\Delta_\perp$) and tangential ($\Delta_\parallel$) components of the displacement vector.

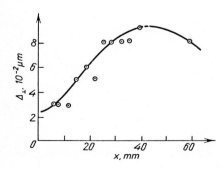

Fig.6.34. Variation of the variance of the displacement vector out-of-plane component along the specimen

7. Holographic Contour Mapping

The macrorelief of a surface is studied by comparing it to a reference surface and looking for deviations. If one uses a plane as a reference, then the lines which are the intersections of the object's surface with a set of equally spaced planes parallel to the reference plane are the contour lines characterizing the relief of the surface. These are called *absolute surface contour lines*.

In comparing reliefs, one can use any arbitrary surface as a reference. For instance, the surface of a manufactured part of any geometry can be compared with that of a reference one. In this case the surface being mapped will be intersected by equidistant reference surfaces, and one observes a difference surface relief and *difference surface contouring*.

With conventional interferometry one can also produce contour maps in both absolute and difference reliefs. To obtain absolute relief maps, it is sufficient to replace one of the plane end mirrors in the Michelson interferometer (Fig.7.1) with the surface under investigation. If we also replace the second plane mirror with a reference surface, however, and match the images of the two surfaces to be compared on the interferogram, the contour map will reproduce a difference relief. The *value of one fringe* (i.e., the spacing between two adjacent intersecting planes) will then be $\lambda/2$, which for the He–Ne laser is 0.315 μm. Thus, this method has a very high sensitivity and is capable of revealing deviations from a reference surface on the order of a few tenths or even hundredths of a micrometer. At the same time, such a high sensitivity imposes limitations on the potential of this method, that is, it cannot be used to study either surfaces with high relief gradients and depths or those with a microstructure comparable in height to the fringe spacing. Indeed, in the first case the number of the contour lines and their frequency will be too high, thus making the fringes unresolvable.

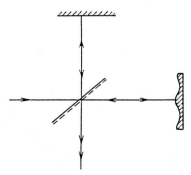

Fig.7.1. Michelson interferometer for absolute surface relief studies

In the second case, the random microstructure of the surface will have a greater influence on the positions and the structures of the contour lines than its macroscopic features.

As discussed in the preceding chapters, while holographic methods do permit investigation of surfaces with microstructure, one actually compares the recordings of shifted or deformed surfaces with identical microscopic features, so that the fringes produced in this way reflect only the macroscopic changes of the surfaces.

The problem of obtaining contour maps of objects with rough surfaces would be solved if one could produce interferograms with light of a wavelength much larger than the microrelief depth. Then, any differences in surface microrelief would not affect the shape of the interference fringes. One of the approaches is to take interferograms with infrared or microwave radiation. Additional technical difficulties are encountered in recording interferograms in the long-wavelength spectral range.

The methods for reducing the sensitivity (i.e., increasing the *value of a fringe*) in low-sensitivity holography are the subject of the present chapter. These methods can potentially be used to measure not only the surface relief but also changes in surface shape in cases where conventional methods of holographic interferometry turn out to be too sensitive. In addition, under certain conditions, application of conventional holographic methods is impossible because the waves scattered by the object in its two states are decorrelated, or because of changes in the object's microstructure, or because of large surface displacements. In such situations the only possible way to study surfaces may be to use holographic topography. This method is useful, for example, for studies of wear of contacting surfaces or in the measurement of large plastic strains, and monitoring shapes of parts.

7.1 Absolute Surface Contour Mapping

7.1.1 Two-Wavelength Method

The two-wavelength method for contouring was proposed by *Hildebrand* and *Haines* [7.1]. The surface is holographed by light of two closely spaced wavelengths λ_1 and λ_2. Such a two-wavelength hologram can be considered as an incoherent superposition of two holograms. When reconstructed, each of these holograms produces an image of the surface in the light of wavelength λ_3 of the reconstructing source.

In general, the difference between the wavelengths used in hologram recording and reconstruction results in a shift of the object image and a change of its scale, which depends on the ratio μ of the wavelengths, i.e., on $\mu_1 = \lambda_3/\lambda_1$ and $\mu_2 = \lambda_3/\lambda_2$. The interference between the simultaneously reconstructed images causes them to be superimposed by fringes which characterize the difference in their reliefs. This is due to the difference in the longitudinal (elevation) magnification. The difference in the transverse magnifications and their transverse shift due to hologram dispersion are hindering factors, however. These factors result in a transverse shift of the corresponding points in the image microstructure, which leads to deformation of the contour lines and to a shift of their localization surface relative to the object surface. These effects are similar in character to those revealed in double-exposure holography of an object when it is rigidly displaced.

One can avoid the transverse magnification of an object and the concomitant undesirable effects by using plane waves for both reference and reconstruction. By (1.12), the transverse magnification $M_\perp$ of a hologram can be written as

$$M_\perp = \frac{1}{1 + \frac{m^2}{\mu}\frac{x_{1P}}{x_{1B}} - \frac{x_{1P}}{x_{1O}}} , \qquad (7.1)$$

where x_{1P}, x_{1B}, x_{1O} are, respectively, the distances from the hologram to the object, the reconstruction, and reference sources (Fig.1.6). For the case of plane reconstruction and reference waves ($x_{1B} = x_{1O} = \infty$), the transverse magnification $M_\perp = 1$.

For the same conditions, for the transverse coordinate of the image (1.10) can be written in the form

$$x_{2r} = x_{2P} + \left[\frac{1}{\mu}\frac{x_{2B}}{x_{1B}} - \frac{x_{2O}}{x_{1O}}\right]x_{1P} . \qquad (7.2)$$

Assuming the reference and reconstruction beams to lie in the $x_1 x_2$ plane (Fig.1.6), we have

$$x_{2r} = x_{2P} + [(1/\mu)\alpha_B - \alpha_0]x_{1P} , \qquad (7.3)$$

where $\alpha_B = x_{2B}/x_{1B}$ and $\alpha_0 = x_{2O}/x_{1O}$ are the angles that the reconstruction and reference beams make with the x_1 axis. (It should be recalled that (1.10) was derived in the small-angle approximation).

To avoid the transverse shift of the reconstructed images with different wavelengths, i.e. to ensure that $x_{2r}(\lambda_1) = x_{2r}(\lambda_2)$, it is sufficient, as follows from (7.3), that

$$\alpha_B/\lambda_3 = \alpha_{01}/\lambda_1 = \alpha_{02}/\lambda_2 . \qquad (7.4)$$

The simplest way to meet this condition is to use Denisyuk's configuration

$$\alpha_B = \alpha_{01} = \alpha_{02} = 0 . \qquad (7.5)$$

An alternative approach to satisfy (7.4) was described by *Varner* [7.2]. He proposed to form the reference beam by means of a diffraction grating, thus, automatically meeting (7.4). By taking into account (7.4) and recording and reconstructing in parallel beams, we can avoid the undesirable transverse displacement of the corresponding points in the reconstructed images of the surface under investigation.

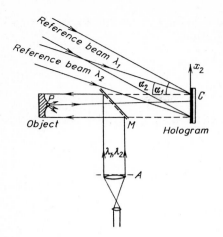

We shall now turn to calculating the path difference of the waves emerging from these points. The character and location of the fringes in the pattern depend on this difference. Consider an arrangement (Fig.7.2) where the object of interest is illuminated with a plane wave containing light of two wavelengths from the side of the hologram by means of a semitransparent mirror M. Let P be a point on the object with coordinates x_{1P}, x_{2P}. The phase will be measured from the plane A (dashed line in the lower part of Fig.7.2) which, just like the hologram, is a distance x_{1P} away from the point P.

For each wavelength, the phase of the illuminating wave at point P will be

$$\psi_{P1} = \frac{2\pi x_{1P}}{\lambda_1}, \quad \psi_{P2} = \frac{2\pi x_{1P}}{\lambda_2}. \tag{7.6}$$

The phases of the spherical waves incident from point P will be

$$\psi_1 = \frac{2\pi x_{1P}}{\lambda_1} + \frac{2\pi r}{\lambda_1}, \quad \psi_2 = \frac{2\pi x_{1P}}{\lambda_2} + \frac{2\pi r}{\lambda_2}, \tag{7.7}$$

where r is the distance from point P to an arbitrary point C on the hologram with the coordinate x_2 measured from the hologram's center:

$$r = \sqrt{x_{1P}{}^2 + (x_2 - x_{2P})^2} \simeq x_{1P}[1 + (x_2 - x_{2P})/2x_{1P}{}^2]. \tag{7.8}$$

In the small-angle approximation, $r = x_{1P}$, so that

$$\psi_1 = \frac{4\pi x_{1P}}{\lambda_1}, \quad \psi_2 = \frac{4\pi x_{1P}}{\lambda_2}. \tag{7.9}$$

We direct two reference waves onto the hologram under the angles α_1 and α_2. For the phases of these waves we can write

$$\psi_{01} = \frac{2\pi x_2 \alpha_1}{\lambda_1}, \quad \psi_{02} = \frac{2\pi x_2 \alpha_2}{\lambda_2}. \tag{7.10}$$

The hologram records the phases of the elementary spherical waves relative to the corresponding reference waves:

$$\psi_1 - \psi_{01} = \frac{4\pi x_{1P}}{\lambda_1} - \frac{2\pi x_2 \alpha_1}{\lambda_1} = \frac{2\pi}{\lambda_1}(2x_{1P} - x_2 \alpha_1),$$

$$\psi_2 - \psi_{02} = \frac{4\pi x_{1P}}{\lambda_2} - \frac{2\pi x_2 \alpha_2}{\lambda_2} = \frac{2\pi}{\lambda_2}(2x_{1P} - x_2 \alpha_2). \tag{7.11}$$

By illuminating the hologram with a plane wave incident under an angle α_3 and with the wavelength λ_3, we reconstruct the original phase distributions of the waves emerging from individual points on the object relative to the reconstruction wave, so that the phases of the reconstructed elementary waves in the hologram plane will be

$$\psi_{1B} = \psi_1 - \psi_{01} + \frac{(2\pi x_2 \alpha_3)}{\lambda_3} = \frac{2\pi}{\lambda_1}(2x_{1P} - x_2 \alpha_1) + \frac{2\pi}{\lambda_3}x_2 \alpha_3,$$

$$\psi_{2B} = \psi_2 - \psi_{02} + \frac{(2\pi x_2 \alpha_3)}{\lambda_3} = \frac{2\pi}{\lambda_2}(2x_{1P} - x_2 \alpha_2) + \frac{2\pi}{\lambda_3}x_2 \alpha_3. \tag{7.12}$$

The phase difference of these waves in the hologram plane is

$$\Delta\psi = \psi_{1B} - \psi_{2B} = 4\pi x_{1P}\left[\frac{1}{\lambda_1} - \frac{1}{\lambda_2}\right] - 2\pi x_2 \left[\frac{\alpha_1}{\lambda_1} - \frac{\alpha_2}{\lambda_2}\right]. \tag{7.13}$$

If condition (7.4) is met, the second term on the right-hand side of (7.13) vanishes, yielding

$$\Delta\psi = 4\pi x_{1P} \frac{\Delta\lambda}{\lambda_1 \lambda_2}, \tag{7.14}$$

so that the phase difference is proportional to the distance of the surface points from the hologram.

This phase difference corresponds to waves incident on the hologram from the point P on the object. Other points of the object with other coordinates x_{1Q} and x_{2Q} also scatter waves to the same point of the hologram. As a result of superposition between numerous pairs of elementary waves with phase differences varying in accordance with the surface relief, we will not be able to see any regular interference pattern in the hologram plane. The waves emerging from different surface points will best be separated from one another at the surface of the object which then represents the localization surface of the interference pattern. For the phase difference, in the small-angle approximation, it is defined by (7.14) for any plane, including the plane of localization.

The spacing between the maxima of the adjacent interference fringes corresponds to a change in the phase difference by 2π so that

$$\Delta x_{1P} = \frac{\lambda_1 \lambda_2}{2\Delta\lambda} \simeq \frac{\lambda^2}{2\Delta\lambda} , \qquad (7.15)$$

which determines the value of one contour line.

A more rigorous analysis, which does not involve either the small-angle approximation or the assumption that the illuminating and reference beams are parallel, yields the conclusion that the contour lines actually represent intersections of the surface under study with a family of ellipsoids of revolution whose common principal axis connects the hologram center with the illumination source. If we place the illumination source at the hologram center, the ellipsoids will degenerate into spheres. As the object is moved away from the hologram and the illumination source, the surfaces of the ellipsoids (or spheres) will flatten out to planes.

In practice, it is important to correctly choose the illumination source so that it provides light of wavelengths corresponding to the given value of the fringe. For this purpose, one can employ different pairs of lines generated by argon and krypton [7.1, 3, 4], dye [7.5], and pulsed ruby [7.6] lasers. In the latter case, when one uses stacks or a Fabry-Perot interferometer with reflecting surfaces separated by the distance d as an exit mirror, the laser will generate a doublet with a wavelength spacing $\Delta\lambda = \lambda^2/2d$. Therefore, for the value of a fringe we obtain

$$\Delta x_{1P} = \frac{\lambda^2}{2\Delta\lambda} = d , \qquad (7.16)$$

which is the thickness of the etalon. Figure 7.3 exhibits a contour map generated in this way by means of a pulsed ruby laser [7.6].

Fig.7.3. Holographic contour maps obtained by two-wavelength method ($\Delta\lambda = 0.125\,\mu m$, $\Delta h = 23\,mm$)

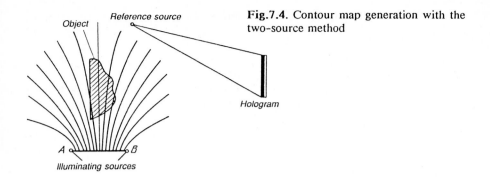

Fig.7.4. Contour map generation with the two-source method

7.1.2 Two-Source Method

This method was first proposed in [7.1] by *Hildebrand* and *Haines* and is illustrated in Fig.7.4. Here A and B are either the sequential positions of a source illuminating the object in two exposures, or the positions of two coherent sources illuminating the object simultaneously in the one-exposure modification. The latter method lends itself more easily to a graphic interpretation. A system of nodes and antinodes shaped as hyperboloids of revolution will form in the space around the two coherent point light-sources. The object placed in this system will be superimposed with contour lines representing intersections of the object with these hyperboloids. In this version of the two-source method, a hologram need not be recorded at all – the contour lines can be studied visually in real time or be photographed.

In a modification of this method the hologram is exposed two times with the light source at position A and then at position B. Since the coordinates of the image do not depend on the position of the illumination source, the two reconstructed images coincide in space and differ only in the phases of the scattered waves. Obviously, the positions of lines in both methods of the two simultaneous sources and the displaced single source are identical.

If the sources A and B lie far from the surface under study or, better still, are moved to infinity by using a collimator, the intersecting surfaces may be approximated with good accuracy by equidistant planes separated by the distance

$$\Delta h = \frac{\lambda}{2\sin(\alpha/2)} \, , \tag{7.17}$$

where α is the angular separation of the sources. The lines generated in this way make absolute contour maps provided that the planes are perpendicular to the viewing direction. The obvious shortcoming of the method is the existence of shadowed areas on the object surface, i.e., areas which cannot be illuminated with sources in the given positions.

The way out of this difficulty is to illuminate the object from many directions. For this purpose, one can use an optical arrangement where,

196

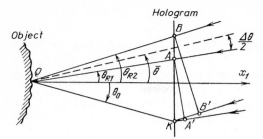

Fig.7.5. Recording a reflection hologram with the two-source method

upon reflection from a rotatable mirror, the light beam is split thus illuminating the object from opposite directions [7.7].

Another approach is to illuminate the object under an acute angle to the viewing direction. This improves the uniformity of illumination and prevents the formation of shadows. The lines obtained under these conditions, while not being equal relief depth contours, can nevertheless be used to calculate such contour maps as has been shown in [7.8].

If the displacement of the light source is accompanied by shifting the reference source through the same angle, then the intersecting surfaces will be parallel to the hologram plane. In Denisyuk's configuration, the same light source produces both the object and reference beams which is the simplest way of providing such synchronous and equal angular displacement [7.9].

Consider the configuration displayed in Fig.7.5. Let Q be an object point separated by the distance x_1 from the hologram. The hologram (and the object) is illuminated by a "plane wave" from a laser. During the first exposure this wave is incident on the hologram at the angle θ_{R1}, and during the second exposure at θ_{R2}. When reconstructing the wavefront, the point Q is viewed through the point K on the hologram, i.e., under an angle θ_0. The angles are measured, as shown in Fig.7.5.

We now calculate the path differences Δ_1 and Δ_2 at point K between the object and reference beams in the first and second exposures:

$$\Delta_1 = AQ + QK - A'K , \qquad (7.18)$$

$$\Delta_2 = BQ + QK - B'K . \qquad (7.19)$$

The change in the path difference is

$$\Delta_1 - \Delta_2 = AQ - BQ - A'K + B'K . \qquad (7.20)$$

Since

$$AQ = x_1/\cos\theta_{R1} , \quad BQ = x_1/\cos\theta_{R2} ,$$

$$AK' = x_1 (\tan\theta_{R1}\sin\theta_{R1} + \tan\theta_0\sin\theta_{R1}) ,$$

$$B'K = x_1 (\tan\theta_{R2}\sin\theta_{R2} + \tan\theta_0\sin\theta_{R2}) ,$$

we obtain

$$\Delta_1 - \Delta_2 = x_1[1/\cos\theta_{R1} - 1/\cos\theta_{R2} - \tan\theta_{R1}\sin\theta_{R1} - \tan\theta_0\sin\theta_{R1}$$

$$+ \tan\theta_{R2}\sin\theta_{R2} + \tan\theta_0\sin\theta_{R2}] . \tag{7.21}$$

Thus, the path difference turns out to be proportional to the distance along x_1 between points on the surface and the hologram and the interference fringes yield a contour map of the relief.

By setting $\Delta_1 - \Delta_2 = \lambda$, one can find the value of one fringe, Δx_1

$$\lambda = \Delta x_1[(1/\cos\theta_{R1} - \sin^2\theta_{R1}/\cos\theta_{R1}) - \tan\theta_0\sin\theta_{R1}$$

$$- (1/\cos\theta_{R2} - \sin^2\theta_{R2}/\cos\theta_{R2}) + \tan\theta_0\sin\theta_{R2}] . \tag{7.22}$$

After some manipulation we eventually arrive at

$$\Delta x_1 = \frac{\lambda\cos\theta_0}{2\sin[(\theta_{R1}+\theta_{R2})/2 + \theta_0]\sin(\theta_{R2}-\theta_{R1})/2} , \tag{7.23}$$

or

$$\Delta x_1 = \frac{\lambda\cos\theta_0}{2\sin(\bar{\theta}-\theta_0)\sin(\Delta\theta/2)} \tag{7.24}$$

where $\bar{\theta} = (\theta_{R1}+\theta_{R2})/2$ is the mean angle of incidence, and $\Delta\theta = (\theta_{R2}-\theta_{R1})/2$ is the change in the angle of incidence. The sensitivity of this method is substantially lower (and the value of a fringe greater) than is the case with the conventional two-source method for the same value of $\Delta\theta$. By varying the viewing direction R_0 one can change the value of a fringe within a fairly broad range.

An example illustrating this method is shown in Fig.7.6a, a holographic topogram of the absolute relief of a specimen tested for fracture toughness. The value of one fringe is 100 μm. The pattern shows an area of con-

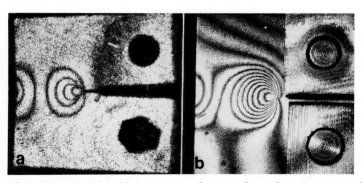

Fig.7.6. (a) Holographic topogram of the surface of a cracked specimen, and (b) interferogram of this surface

198

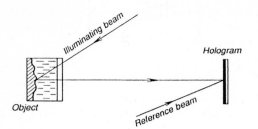

Fig.7.7. Schematic diagram of the immersion method

traction near the tip of the crack. For comparison, Fig.7.6b displays a conventional two-exposure holographic interferogram showing the buildup of residual deformations of the same specimen obtained at substantially lower loading levels (the value of a fringe is $0.31\,\mu$m). This pattern was recorded by the technique described in Sect.6.1.1.

7.1.3 Immersion Method

In the two-exposure immersion method for producing contour maps according to *Tsuruta* et al. [7.10] the object under study is placed into an immersion cell (Fig.7.7) with a plane window filled before the first exposure with a liquid or gas with refractive index n_1, and before the second exposure, with that of a refractive index n_2. As will be shown below, this is equivalent to using an effective wavelength $\lambda_0 = \lambda/(n_1 - n_2)$ which permits varying the value of a fringe within a broad range.

If the plane of the cell window is normal to the direction of the propagation of the object wave, then a change in the refractive index will not significantly affect the angle of incidence of the object beam on the hologram. Since neither the wavelength nor the incidence angle of the reference beam change under these conditions, there will be no lateral displacement and no change in the scale of the reconstructed image, thus automatically implying that the fringes will localize on the object surface.

We calculate the path difference for two beams arriving at the same point B of an object surface and being reflected from it in a direction perpendicular to the window of the immersion cell (Fig.7.8). As seen in the figure,

$$\delta\ell = (CB+h)n_1 - (AB+h)n_2 - DE . \tag{7.25}$$

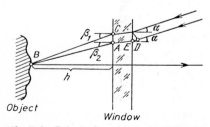

Fig.7.8. Calculating the path difference in the immersion method

199

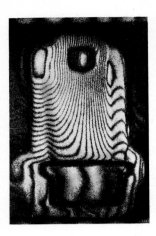

Fig.7.9. Contour map of a turbine blade die obtained by the immersion method (ethylene glycol-water mixture, $\Delta h = 2$ mm)

Since $CB = h/\cos\beta_1$, $AB = h/\cos\beta_2$, $DE = h(\tan\beta_1 - \tan\beta_2)\sin\alpha$, and also noting that $\sin\alpha = n_1\sin\beta_1 = n_2\sin\beta_2$, we obtain after some manipulation

$$\delta\ell = [(1+\cos\beta_1)n_1 - (1+\cos\beta_2)n_2]h . \tag{7.26}$$

The value of a fringe corresponds to a change of δl in one wavelength, i.e.,

$$\Delta h = \lambda/[(1+\cos\beta_1)n_1 - (1+\cos\beta_2)n_2] . \tag{7.27}$$

When illuminating the object from the side of the hologram ($\cos\beta_1 = \cos\beta_2 = 1$) in a configuration similar to that in Fig.7.2, (7.27) transforms into

$$\Delta h = \frac{\lambda}{2(n_1 - n_2)} . \tag{7.28}$$

Tsuruta et al. [7.10] used freon as the immersion gas which permits $\Delta h = 0.3$ mm when the pressure is changed by 10 MPa. Mixtures of water (n = 1.333) and ethylene glycol (n = 1.427), as well as water and alcohol (n = 1.362) also enable the variation of Δh over a broad range. Figure 7.9 displays a contour map generated in this way ($\Delta h = 2$ mm) of a die for stamping of turbine blades [7.11]. The immersion method has also been used in wear studies [7.12].

7.2 Difference Surface Contour Mapping

As already pointed out, it is quite frequently desirable to compare the macrorelief of a surface under study with a reference one, i.e., to generate difference contour maps. In this case the contour lines are the intersections of the surface in question with equidistant surfaces of the reference. Contour maps in difference relief can naturally be obtained also by calculations

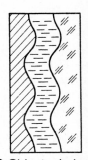

a Object window

Fig.7.10. (a) Schematic diagram of immersion cell with reference window, (b) the specimen, (c) difference, and (d) absolute surface relief contour maps

based on the absolute elevation contour maps of the surfaces being compared, i.e., by subtracting the elevations of both surfaces at each point. Obviously, this procedure is too tedious and too inaccurate since it involves finding small differences between two comparatively large elevations. In this section we consider a number of holographic techniques which can be used for a straightforward generation of difference-relief contour maps.

7.2.1 Method of the Cell Reference Window

What one actually measures in the conventional immersion method is the thickness h of the liquid layer between the inner plane surface of the cell window and the surface of the object under study (Fig.7.7). Thus, if we fabricate a cell whose transparent window has an inner surface with a relief which is inverted relative to that of the etalon (Fig.7.10a), then the thickness of the liquid layer will characterize the difference relief [7.13]. In doing this, one must, naturally, reduce the light refraction at the curved interface between the cell window and the immersion liquid to a minimum. Therefore, the refractive index of the cell window should be as close as possible to that of the liquids (for numerical estimates, see [7.14]). In addition, when mounting the specimen in the cell, one must minimize both its tilt and transverse displacement relative to the etalon window. If the first condition is not met, finite-width fringes due to the liquid wedge will appear, while in the second case the path difference will be determined not only by a difference in the relief, but also by the gradient in the surface relief. (This arrangement can be used to generate *surface gradient contour maps* where the surface under study is displaced in the transverse direction with respect to its own transparent replica which serves as the cell window).

The above method of difference relief measurement can be illustrated by a study of the surface profile over a cylindrical cut made in a cylinder (Fig.7.10b). Figure 7.10c presents the difference contour map obtained with a reference window of cylindrical shape. Interference fringes are observed only on the surface of the recess. For comparison, Fig.7.10d shows an absolute contour map of the same object with interference fringes present over the whole surface. The value of a fringe is the same (about 0.5mm) in the two interferograms.

The reference window is an exact replica of the cylinder and is produced by polymerization of polyester resin (n = 1.54) in a mould [7.15]. Special precautions were taken to avoid adhesion between the resin and the object surface. Distilled water (n = 1.33) and an aqueous solution of glycerine (n = 10^{-3}) were used as immersion liquids. Analysis shows that the error in difference relief measurement due to the difference in the refractive indices of the cell window and liquid do not exceed a fraction of one percent. Such a small error (despite the large difference in the refractive indices) originates from small relief gradients of the surface under study.

7.2.2 Method of the Etalon Reference Wave

Another way of obtaining a difference contour map is to use a holographic setup where the object and the reference waves are formed in the same manner, namely, by reflection from the surface under study and an etalon surface [7.16]. Both objects are placed in an immersion cell and, just as in the conventional immersion method, the value of a fringe can be found from (7.28). Now, however, it is the value of a difference-relief contour line.

The arrangement is shown in Fig.7.11. The images of the objects focussed onto a photographic plate are holographed in the opposed-beam configuration. Reconstruction is made in white light with a beam of arbitrary shape. (It may be recalled that an image hologram permits reconstruction of undistorted images irrespective of the structure of the reconstruction wave). This method, however, has the shortcoming in that both the object and the reference waves have speckles [7.17], and the shift of the speckle patterns due to a change in the refractive index of the immersion liquid must be much smaller than the mean speckle size. We have in mind primarily the longitudinal speckle shift and the corresponding condition

$$2h\Delta n \ll 4\lambda \ell^2 /a^2 . \tag{7.29}$$

Here, h is the thickness of the liquid layer between the object and the inner surface of the cell window, Δn is the change of the refractive index, a is

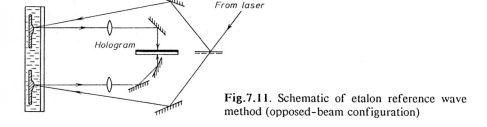

Fig.7.11. Schematic of etalon reference wave method (opposed-beam configuration)

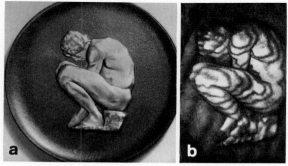

Fig.7.12. (a) A medal and (b) its contour map in absolute relief obtained with arrangement of Fig.7.11. (The etalon was a diffusely scattering metallic plate)

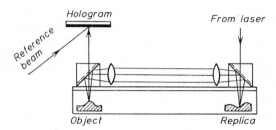

Fig.7.13. Generation of difference relief contour map with the object illuminated by a beam reflected from a replica of an etalon object

the objective lens radius, ℓ is the object-to-lens distance, and $4\lambda\ell^2/a^2$ is the mean longitudinal speckle size. If condition (7.29) is not met, then a shift of the speckles between exposures will tend to degrade the fringe contrast because of decorrelation. When the images are sharply focussed, however, there is practically no transverse speckle shift [7.17].

Holograms obtained by the above method have a poorer diffraction efficiency compared to the conventional types, since, because of the speckles in the object and reference waves, the interference pattern occupies only a fraction of the total volume of the photosensitive medium. An example of a contour map of a metallic medal (Fig.7.12a) generated in this way is presented in Fig.7.12b.

7.2.3 Compensation Method

The arrangement shown in Fig.7.13 shows the optical setup for the compensation method. The object under study and a replica of a reference piece with an inversed relief are placed into an immersion cell. The light from a laser illuminates the replica and upon reflection is projected through an objective lens onto the object surface with unit magnification. If the replica is absolutely exact, then the plane wave from the laser, upon double reflection from the object and its replica, will retain the plane wavefront (we disregard here the speckle structure caused by the microrelief of both surfaces). A two-exposure hologram constructed with immersion liquids of different refractive indices will not reveal any interference fringes. If, however, there are differences in the reliefs of the object and of the rep-

203

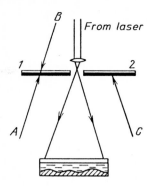

Fig.7.14. Schematic diagram of compensation method

lica, then the wavefront distortions will not be fully compensated, and we will obtain a difference relief contour map. The value of a fringe is then determined, as before, by (7.28). The drawback of this method is that as a result of double reflection from the diffuse surfaces, double passage through the beam splitters and double reflections, the light losses may become very high. Projecting the image of a replica on the object under study is also not a simple task.

Similar compensation can be obtained by using the real image of the reference surface (a conjugate wave) [7.19] for the illumination of the surface. The corresponding arrangement is presented in Fig.7.14. First, the reference object is placed into the immersion cell and hologram *1* is constructed with a reference wave A. Next, after returning the developed hologram *1* back into position (or after in situ developing), it is illuminated with a wave B conjugate to the wave A. As we already know (Chap.1), the reconstructed wave, which is conjugate to the object wave, forms a nondistorted real image of the reference surface on the same surface. If we now replace the reference object with the one under study and mount the latter in exactly the same position as the reference, the situation will be similar to the one in Fig.7.13. We now doubly-expose hologram *2* with different values of the refractive index of the cell liquid to obtain the decompensation fringes for the difference relief contour map.

8. Holographic Studies of Vibrations

As discussed in Chaps.1 and 2, the object cannot, generally speaking, be allowed to move during the recording of a hologram. However, in certain cases (Sect.2.1.5) one can holograph moving objects. The hologram thus obtained provides information on the character of the object's motion and changes in the object, in particular, on deformations suffered in the course of motion.

The simplest case is schematically illustrated in Fig.8.1. The object is a diffusely reflecting plate carrying out harmonic angular oscillations with an amplitude $\alpha_0/2$ during the exposure. Let the hologram exposure time be much larger than the oscillation period. Under these conditions, the hologram pattern will be determined by the value of $\cos(\phi-\psi)$, averaged over the exposure time, as described by

$$I \sim a_0{}^2 + a_r{}^2 + 2a_0 a_r \cos(\phi-\psi) \; ,$$

where a_0 and a_r, ϕ and ψ, are the amplitudes and phases of the object and reference waves, respectively. The points on the object move with various velocities in different sections of their travel, moving most slowly near the two extreme positions of maximum displacement. These two positions make the biggest contribution to the holographic pattern. Thus, each oscillation period of the object can be divided into three parts (Fig.8.2): *1* - the time that the object spends in one extreme position; *2* - the same time for the second extreme position; and *3* - the time of fast motion. We define *1* and *2* to be the time intervals during which the path difference Δ changes by less than $\lambda/4$.

During the times *1* and *2* the object is exposed in its extreme positions yielding an interference pattern similar to that obtained in the two-exposure method. The third part of the exposure during time *3*, rather than con-

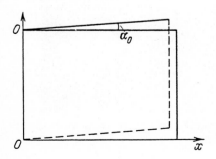

Fig.8.1. Rotation of a plate

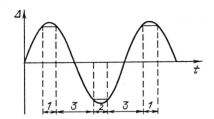

tributing to the holographic pattern, only reduces its contrast. Therefore, the contrast becomes higher the longer the object stays in the extreme positions. Obviously, the "useful" fraction of the oscillation period increases with decreasing amplitude of the oscillations. This is why the bright-fringe intensity decreases with increasing oscillation amplitude, although this does not occur as fast as is the case of a plate rotating at a constant rate (Fig.2.7).

This method was proposed in 1965 by *Powell* and *Stetson* [8.1] in a pioneering work on holographic interferometry. The reconstructed image of the object is here modulated by interference fringes which connect the points oscillating with the same amplitude. Fixed parts on the object (*nodal lines*) will exhibit maximum brightness since, during the entire exposure, these points of the object surface have occupied the same position. The points for which the path difference of the waves scattered by the object in its maximum displacement positions makes up an odd number of halfwaves will be the least intense and thus correspond to the centers of the dark fringes. The points for which this path difference constitutes an even number of halfwaves, will form the maxima of the bright fringes. The intensity of these maxima, however, will decrease with increasing oscillation amplitude, since the time the object spends in its extreme positions reduces

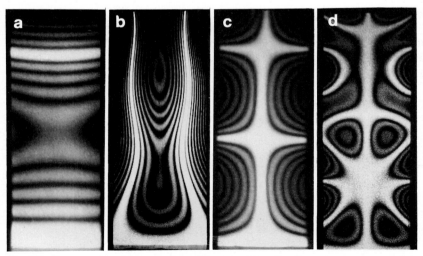

Fig.8.3. Reconstructed image of a clamped plate vibrating at a resonance frequency of: (**a**) 559, (**b**) 2510, (**c**) 2843, and (**d**) 12057 Hz

206

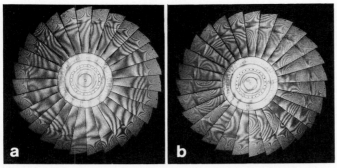

Fig.8.4. Reconstructed image of a cast disc with blades $\phi = 1500$ mm vibrating at resonance frequences of (**a**) 545 Hz and (**b**) 580 Hz. (Courtesy of D.S. Elinevskii, R.S. Bekbulatov and Yu.N. Shaposhnikov)

as the amplitude increases. Thus, the reconstructed image of a vibrating object is overlaid with interference fringes, the brightest of them lying along the nodal line, each of the subsequent lines with decreasing intensity connecting the object points oscillating with the same amplitude (Figs. 8.3,4). A more rigorous analysis supports this purely qualitative reasoning.

8.1 Vibration Analysis of Fixed Objects

8.1.1 The Method of Powell and Stetson

We consider the time-average method of Powell and Stetson in more detail. The complex amplitude of the wave scattered by a point Q on a fixed object and reaching a point B on the hologram can be written as $a_0 \times \exp(-i\phi_0)$, where $\phi_0 = 2\pi x(\sin\theta)/\lambda$. Let the point Q move in the **v** direction with a velocity $|\mathbf{v}|$ (Fig.8.5). Then the phase of the wave scattered by Q to the same point on the hologram depends on time as

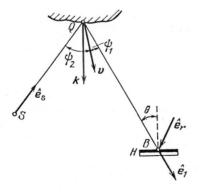

Fig.8.5. Holographing a moving surface

$$\phi = \phi_0 + \frac{2\pi}{\lambda} \int_0^t \mathbf{k} \cdot \mathbf{v} \, dt = \phi_0 + \frac{2\pi}{\lambda} \int_0^t (\cos\psi_1 + \cos\psi_2) v \, dt \,, \tag{8.1}$$

where $\mathbf{k}$ is the sensitivity vector, and the angles ψ_1 and ψ_2 are defined in Fig.8.5. For the complex amplitude we can write $a_0 \exp(-i\phi)$. Its mean value over the exposure interval τ in the approximation of linear recording determines the brightness of the point Q in the reconstructed image of the object:

$$I \sim \left| \frac{a_0}{\tau} \int_0^\tau \exp(-i\phi) \, dt \right|^2$$

$$= \left| \frac{a_0}{\tau} \int_0^\tau \exp\left\{-i\left[\phi_0 + \frac{2\pi}{\lambda} \int_0^t (\cos\psi_1 + \cos\psi_2) v \, dt\right]\right\} dt \right|^2 . \tag{8.2}$$

This expression covers all cases of motion of the point Q during exposure. Some of these cases, namely, translation of an object with a constant velocity and stepwise motion, have been discussed in Sect.2.1.5.

We now consider the harmonic oscillation of the object points when one can write

$$\int_0^t (\cos\psi_1 + \cos\psi_2) v \, dt = A(\cos\psi_1 + \cos\psi_2)\sin\omega t \,, \tag{8.3}$$

where A is the amplitude of oscillation. The general expression (8.2) will then assume the form

$$I \sim \left| \frac{a_0}{\tau} \int_0^\tau \exp\left\{-i\left[\frac{2\pi A}{\lambda}(\cos\psi_1 + \cos\psi_2)\sin\omega t\right]\right\} dt \right|^2 . \tag{8.4}$$

Since

$$\exp\left\{-i \frac{2\pi A}{\lambda}(\cos\psi_1 + \cos\psi_2)\sin\omega t\right\} = \sum_{n=-\infty}^{\infty} J_n\left[\frac{2\pi A}{\lambda}(\cos\psi_1 + \cos\psi_2)\right] e^{in\omega t} \,, \tag{8.5}$$

where J_n is the Bessel function of the first kind of order n, we arrive at

$$I \sim \left| \frac{a_0}{\tau} \sum_{n=-\infty}^{\infty} J_n\left(\frac{2\pi A}{\lambda}(\cos\psi_1 + \cos\psi_2)\right) \int_0^\tau \exp(in\omega t) \, dt \right|^2 \tag{8.6}$$

208

and, after integration, obtain

$$I \sim \left| a_0 \sum_{n=-\infty}^{\infty} J_n \left[\frac{2\pi A}{\lambda}(\cos\psi_1 + \cos\psi_2) \right] \frac{\exp(in\omega t) - 1}{in\omega t} \right|^2 . \qquad (8.7)$$

If the exposure time constitutes an even number of oscillation periods, i.e., $\tau = 2k\pi/\omega$, all terms except the one corresponding to n = 0 will vanish, so that

$$I \sim a_0^2 J_0^2 \left[\frac{2\pi A}{\lambda}(\cos\psi_1 + \cos\psi_2) \right] . \qquad (8.8)$$

Likewise for $\tau \gg 2\pi/\omega$, the zero term in the sum dominates since the factor $[\exp(in\omega\tau)-1]/in\omega\tau$ vanishes as $\omega\tau$ increases, and the distribution of irradiance over the surface of the object is expressed by (8.8), too. Actually, it is easy to understand that a transition in exposure time from $\tau = 2k\pi/\omega$ to $\tau = 2(k+\Delta k)\pi/\omega$, where k is a large number and $\Delta k < 1$, should practically not affect the hologram and, hence, should not change the brightness of the reconstructed image.

For $\psi_1 = \psi_2 = 0$ we rewrite (8.8) in the form

$$I \simeq a_0^2 J_0^2 (4\pi A/\lambda) . \qquad (8.9)$$

Being an oscillating function (Fig.2.7), the relationship (8.9) modulates the brightness of the reconstructed image and thus describes the observed interference fringes. The amplitudes for the first 15 dark fringes obtained for the laser wavelengths $\lambda_1 = 0.633$ μm and $\lambda_2 = 0.514$ μm, which are most widely used in holographic vibration studies, are listed in Table 8.1.

The physical reason behind the decrease in brightness of the fringes corresponding to the regions of the object with large vibrational amplitudes

Table 8.1. Vibration amplitudes corresponding to dark interference fringes for $\lambda_1 = 0.633$ μm and $\lambda_2 = 0.514$ μm ($\psi_1 = \psi_2 = 0$)

n	1	2	3	4	5	6	7	8
$A(4\pi/\lambda)$	2.405	5.520	8.654	11.792	14.931	18.071	21.212	24.352
$A[\mu m](\lambda_1)$	0.121	0.278	0.436	0.594	0.752	0.910	1.069	1.227
$A[\mu m](\lambda_2)$	0.098	0.226	0.354	0.482	0.611	0.739	0.868	0.996
	9	10	11	12	13	14	15	
	27.493	30.635	33.776	36.917	40.058	43.200	46.341	
	1.385	1.543	1.701	1.860	2.018	2.176	2.334	
	1.125	1.253	1.382	1.510	1.638	1.767	1.895	

is as follows. The light scattered by the surface of a vibrating object contains, apart from the original laser frequency ν, also the frequency components $\nu \pm n\omega/2\pi$, where n is an integer, and ω is the circular frequency of the object's vibrations. The larger the vibrational amplitude, the smaller the fraction of scattered light with original frequency ν ($n\neq0$). This is readily seen from (8.9), since the Bessel function of zero[th] order J_0 decreases rapidly with increasing vibrational amplitude A. The region of the object surface with A=0 contributes to the object wave only at the original frequency ν, so that all the waves scattered by stationary regions of the object interfere with the reference. For the vibrating parts the useful contribution to the object wave is much smaller. Indeed, if, for instance, $2\pi A/\lambda = 5$, then the useful contribution constitutes only 2% of the total irradiance.

The instantaneous interference pattern produced by the nth harmonic of an object wave with a reference wave of one frequency ν moves during the exposure with a velocity determined by the difference frequency $n\omega$ and, in accordance with the condition of exposure $\tau \gg 2\pi/\omega$, does not remain recorded in the photographic material as a hologram. Therefore, the components of the object wave with the frequency $\nu\pm n\omega/2\pi$ ($n\neq0$) are recorded as an additional incoherent background. Thus, in studies of objects with large vibrational amplitudes (A$\gg\lambda$) the hologram exposure conditions will be optimum when the intensity of the object wave is equal to, or even far exceeding, that of [8.2]

As follows from the above consideration, the method of *Powell* and *Stetson* reduces to holographing the vibrating object during an exposure time much longer than the vibrational period, and to analyzing the object's vibrations from the interference pattern thus produced. The simplicity of the method, the graphic character of the results obtained, and the wealth and significance of the information gained on vibrational behavior have initiated a large number of studies devoted to the application and development of this method [8.3-15]. Some of these publications, e.g., [8.6, 8, 11, 12, 14, 15] deal with the study of vibrations of relatively simple objects such as thin membranes, diaphragms, plates and beams. Holographic vibration analysis has also been employed to investigate the vibrational characteristics of industrial components of complex shape, however, such as turbine blades and discs [8.4, 5, 7.10, 13, 16].

The time-average technique is applicable to an investigation of multi-mode vibration. Consider an object simultaneously executing two different vibrational motions in two irrationally related modes. In this case, the interferogram will represent a superposition of the fringes due to each separate fundamental frequency. We then have the following expression for the fringe function, in the case of sinusoidal vibrations [8.12, 17]:

$$|M(\delta)|^2 = J_0^2(\delta_1)J_0^2(\delta_2) ,$$

where δ_1 and δ_2 are the functions characterizing the interference fringes for each of the vibrational modes.

For an object vibrating simultaneously in two sinusoidal modes of the same frequency, the fringe function has the form [8.14]

$$|M(\delta)|^2 = J_0^2(\delta_1+\delta_2).$$

Stetson and *Powell* also proposed a method for studying the vibrations of diffuse objects in real time [8.18]. However, the visibility (contrast) of the fringes is substantially lower than that which is typical in the time-average technique. Such an experiment can be performed in the following way. First, a hologram of a fixed object is recorded and either processed in situ or placed back exactly in the original position. Next, similar to the way this is done in the real-time method, the object and the hologram are illuminated, and the vibrating object is observed through the hologram. Behind the hologram, the wave from the vibrating object and the wave from the fixed object, which has been reconstructed by the hologram, will propagate simultaneously. The instantaneous pattern produced by the interference of these waves will be averaged in the course of recording. Thus, in contrast to the averaging of the complex amplitude occurring in the time-average technique, in the real-time method the resultant irradiance is averaged.

Let the complex amplitude of the wave from a fixed object point Q recorded on and reconstructed by the hologram be $-a_0 \exp(-i\phi_0)$ at the point of observation. The minus sign comes from the fact that in the conventional negative recording the phase of the reconstructed wave becomes shifted by π relative to the object wave. The complex amplitude of the wave emerging from the same point of the vibrating object will be $a_0 \exp(-i\phi)$, where ϕ, just as before in (8.1), is determined by $\phi = \phi_0+(2\pi/\lambda)A(\cos\psi_1+\cos\psi_2)\sin\omega t$. The resultant complex amplitude will be $a_0[\exp(-i\phi)-\exp(-i\phi_0)]$, and the instantaneous irradiance

$$I = a_0^2 \left|[\exp(-i\phi) - \exp(-i\phi_0)]\right|^2 = 2a_0^2[1 - \cos(\phi-\phi_0)] . \tag{8.10}$$

For the irradiance averaged over the vibrational period, we have for the quantity detected by the eye or by a photographic plate

$$\langle I \rangle = 2a_0^2 \frac{1}{T} \int_0^T [1 - \cos(\phi-\phi_0)]dt . \tag{8.11}$$

Substituting $\phi-\phi_0$ from (8.1) yields

$$\langle I \rangle = 2a_0^2 \left[1 - J_0 \left(\frac{2\pi}{\lambda}A(\cos\psi_1-\cos\psi_2)\right)\right] , \tag{8.12}$$

which for the case $\psi_1 = \psi_2 = 0$ reduces to

$$\langle I \rangle = 2a_0^2 [1 - J_0(4\pi A/\lambda)] . \tag{8.13}$$

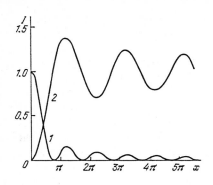

Fig.8.6. Intensity distribution function: (*1*) $J_o^2(x)$ in the time-average method; (*2*) $[1-J_o(x)]$ in the real-time method

A comparison of this expression with (8.9) derived for the time-average technique (Fig.8.6) suggests that the contrast of the fringes observed on the surface of a vibrating object in real-time holography is substantially less than unity, while the fringe period is twice as long.

The real-time method can be modified by tilting the hologram (or the object) before observation, thus producing a finite-width fringe pattern. In this pattern, the regions of high contrast correspond to nodal lines. Figure 8.7 illustrates such a study of the vibration of a plate with two holes, which is clamped on the edges. Figures 8.7a and b present time-average holograms obtained at two close frequencies. Even a slight deviation in the frequency changes the position of the nodes and antinodes. In Fig.8.7c a real-time fringe pattern obtained at the same frequency as the one in Fig.8.7a is shown (the nodes lie on the straight line connecting the hole centers).

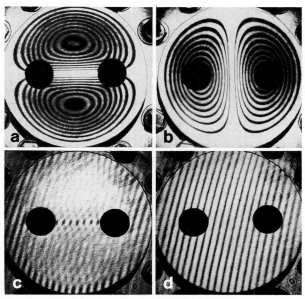

Fig.8.7. Interferograms of a plate vibrating at (**a,c**) 2123 Hz and (**b**) 1974 Hz, obtained by (**a,b**) time-average, and (**c**) real-time technique; (**d**) carrier fringe interferogram

212

Figure 8.7d displays a real-time interferogram corresponding to a stationary plate. One can readily see the carrier fringes whose frequency and orientation are determined by the tilt of the plate.

Despite its inherent simplicity and the possibility of its application to diverse problems of the vibration analysis, the method of *Powell* and *Stetson* also possesses a number of shortcomings:

1) Although all fringes have the maximum possible contrast, their brightness drops rapidly with the increasing amplitude; indeed, the tenth fringe is 2% as bright as the nodal line, and the twentieth only 1%. Therefore, studies of objects with a vibration amplitude in excess of 5λ are practically impossible.

2) While this method permits, in principle, real-time studies of vibrations, the fringes obtained under these conditions are of low contrast, particularly for large amplitudes.

3) The method does not provide objective information on the relative phases of vibration of different points on the object surface.

4) Analysis of very small amplitude vibrations ($A < \lambda/4$) cannot be carried out.

The first three drawbacks can be eliminated by using the stroboscopic holographic method. The fourth and, partially, the first shortcomings can be counteracted by phase-modulating the reference beam and employing the so-called holographic subtraction to be discussed below.

8.1.2 Stroboscopic Holographic Interferometry

By stroboscopic holography we define holographic methods for studying repetitive processes which involve the exposure of holograms to multiply repeated light pulses that are synchronized with a certain phase of the process. This method was first proposed by *Zaidel* and *Ostrovsky* [8.19] in 1967 to study ac-powered gas plasmas. However, stroboscopic holography turned out to be best suited to the vibration analysis of opaque objects [8.20-24].

We consider the theory underlying the stroboscopic holographic method, limiting ourselves to the case where the object is illuminated from the side of the hologram ($\psi_1 = \psi_2 = 0$) and is vibrating in the same direction. The complex amplitude of the object wave incident on the hologram from an object point Q with an amplitude A is

$$a_0 \exp\{-i[\phi_0 + (4\pi/\lambda) A \sin\omega t]\}.$$

The intensity of the reconstructed image is proportional to the squared modulus of this quantity, averaged over the exposure time τ,

$$I = \left| \frac{a_0 \exp(-i\phi_0)}{\tau} \int_0^T \exp\left(-i\frac{4\pi}{\lambda} A \sin\omega t\right) dt \right|^2 . \tag{8.14}$$

Expanding the integrand in a series in Bessel functions we find

$$I = \left| \frac{a_0}{\tau} \sum_{n=-\infty}^{\infty} J_n(4\pi A/\lambda) \int_0^{\tau} \exp(in\omega t) dt \right|^2 . \qquad (8.15)$$

Assuming the strobe pulse length to be equal to a kth part of the vibration period, i.e., $\tau = 2\pi/\omega k = T/k$ and synchronizing it with one of the two maximum object displacement positions we obtain

$$I = a_0^2 \left| \frac{k}{T} \sum_{n=-\infty}^{\infty} J_n\left(\frac{4\pi}{\lambda}A\right) \int_{\frac{T}{4} - \frac{T}{2k}}^{\frac{T}{4} + \frac{T}{2k}} \exp\left(in\frac{2\pi t}{T}\right) dt \right|^2 . \qquad (8.16)$$

The integral is

$$\frac{T}{k}\left(\sin\frac{n\pi}{k} \Big/ \frac{n\pi}{k}\right) \exp\frac{in\pi}{2}$$

so that

$$I = a_0^2 \left| \sum_{n=-\infty}^{\infty} J_n\left(\frac{4\pi}{\lambda}A\right) \frac{\sin(n\pi/k)}{n\pi/k} \exp\left(\frac{in\pi}{2}\right) \right|^2 . \qquad (8.17)$$

Since $J_{-n} = (-1)^n J_n$ and $\exp(in\pi/2) = i^n$ we have

$$J_n \exp(in\pi/2) = J_{-n} \exp(-in\pi/2),$$

$$I = a_0^2 \left| J_0\left(\frac{4\pi}{\lambda}A\right) + 2 \sum_{n=1}^{\infty} J_n\left(\frac{4\pi}{\lambda}A\right) \frac{\sin(n\pi/k)}{n\pi/k} i^n \right|^2 . \qquad (8.18)$$

Similarly, for the case of illumination of an image reconstructed with a hologram exposed to pulses of the same duration, T/k, but synchronized with the other maximum displacement position of the object we obtain

$$I = a_0^2 \left| \frac{k}{2T} \sum_{n=-\infty}^{\infty} J_n\left(\frac{4\pi}{\lambda}A\right) \int_{\frac{3T}{4} - \frac{T}{2k}}^{\frac{3T}{4} + \frac{T}{2k}} \exp\left(in\frac{2\pi t}{T}\right) dt \right|^2$$

$$= a_0^2 \left| J_0\left(\frac{4\pi}{\lambda}A\right) + 2 \sum_{n=1}^{\infty} J_n\left(\frac{4\pi}{\lambda}A\right) \frac{\sin(n\pi/k)}{n\pi/k} (-i)^n \right|^2 . \qquad (8.19)$$

214

If we expose a hologram in the two extreme positions of the object, both images described by (8.18 and 19) will be reconstructed, yielding an interference pattern formed by their superposition with the following irradiance distribution:

$$I = a_0^2 \left| \frac{k}{T} \sum_{n=-\infty}^{\infty} J_n\left(\frac{4\pi}{\lambda}A\right) \left\{ \int_{\frac{T}{4}-\frac{T}{2k}}^{\frac{T}{4}+\frac{T}{2k}} \exp\left(in\frac{2\pi t}{T}\right) dt \right. \right.$$

$$\left. \left. + \int_{\frac{3T}{4}-\frac{T}{2k}}^{\frac{3T}{4}+\frac{T}{2k}} \exp\left(in\frac{2\pi t}{T}\right) dt \right\} \right|^2$$

$$= a_0^2 \left| J_0\left(\frac{4\pi}{\lambda}A\right) + \sum_{n=1}^{\infty} J_n\left(\frac{4\pi A}{\lambda}\right) \frac{\sin(n\pi/k)}{n\pi/k} [i^n + (-i)^n] \right|^2 . \qquad (8.20)$$

The quantity in square brackets is $2i^n$ for even n and zero for odd n. Accordingly, the terms in the sum with the odd index will drop out and the interference pattern takes on the following irradiance distribution:

$$I = a_0^2 \left| J_0\left(\frac{4\pi}{\lambda}A\right) + 2 \sum_{p=1}^{\infty} J_{2p}\left(\frac{4\pi}{\lambda}A\right) \frac{\sin(2p\pi/k)}{2p\pi/k}(-1)^p \right|^2 . \qquad (8.21)$$

One readily sees that in the particular case k = 2 (continuous exposure of the object) $\sin(2p\pi/k) = 0$ for any p, yielding the relation of Powell and Stetson

$$I = a_0^2 J_0^2\left(\frac{4\pi}{\lambda}A\right) . \qquad (8.22)$$

When k→∞ we get the result of the two-exposure method. Indeed, substituting $\cos p\pi$ for $(-1)^p$ in (8.21) and assuming $\sin(2p\pi/k)/(2p\pi/k) = 1$ we have

$$I = a_0^2 \left| J_0\left(\frac{4\pi}{\lambda}A\right) + 2 \sum_{p=1}^{\infty} J_{2p}\left(\frac{4\pi}{\lambda}A\right) \cos 2p\frac{\pi}{2} \right|^2 , \qquad (8.23)$$

or, applying the well-known property of the Bessel functions with even index we obtain

$$I = a_0^2 \cos^2(4\pi A/\lambda) . \qquad (8.24)$$

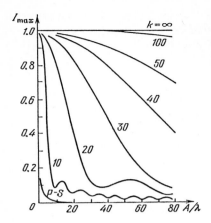

Fig.8.8. Fringe maximum intensity vs. vibration amplitude for the time-average method (P-S) and stroboscopic holography for different stroboscopic pulse duty cycles k (for the time-average technique, k = 2)

Indeed, this is the distribution of irradiance in the two-exposure method.

Figure 8.8 presents the maximum fringe irradiance calculated by (8.21,22 and 24) as a function of the vibration amplitude [Eq.(8.21) for different k]. As seen in the figure, the irradiance at the fringe maxima in the stroboscopic method falls off much slower than it does in the Powell-Stetson technique so that even at duty factors of $k \simeq 10$ to 20 one can study vibrations with amplitudes of a few tens of wavelengths.

Consider the case where the strobe pulse is synchronized with an arbitrary rather than the extreme position of an object. Obviously, the requirements imposed on the pulse duration will be least rigorous for the conditions where the object is holographed close to the rest position. If, however, the vibration amplitude is so high that the number of fringes in the reconstructed image becomes too large to resolve, the strobe pulse can be shifted in time to reduce the sensitivity. Such experiments were carried out by *Zaidel* et al. [8.20].

We take, as before, the duration of the strobe pulse to be T/k; but now it will be centered at an arbitrary moment of time αT (or $\alpha T + T/2$) rather than at T/4 (or 3/4T). For $\alpha = 1/4$ this case reduces to the previously considered one. Thus, we must find the function (8.16) for new, generalized, limits of integration, from $\alpha T - T/2k$ to $\alpha T + T/2k$, i.e.,

$$ I = a_0^2 \left| \frac{k}{T} \sum_{n=-\infty}^{\infty} J_n\left(\frac{4\pi}{\lambda}A\right) \int_{\alpha T - \frac{T}{2k}}^{\alpha T + \frac{T}{2k}} \exp\left(in\frac{2\pi t}{T}\right) dt \right|^2 . \qquad (8.25) $$

Upon integration we obtain

$$ I = a_0^2 \left| \sum_{n=-\infty}^{\infty} J_n\left(\frac{4\pi}{\lambda}A\right) \frac{\sin(n\pi/k)}{n\pi/k} \exp(in2\pi\alpha) \right|^2 . \qquad (8.26) $$

216

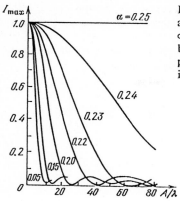

Fig.8.9. Fringe maximum intensity vs. vibration amplitude with the stroboscopic pulse, shifted by αT from the moment of transit through the equilibrium for different values of α ($\alpha = 0.25$ corresponds to the maximum displacement of the vibrating object). The duty cycle k = 100

The function (8.26) is plotted in Fig.8.9 for k = 100 and for different values of α. As seen from the plots, a shift of the strobing moment by only a few percent of the vibration period results in a drastic reduction of fringe intensity, particularly in higher orders. If, however, the strobing duty factor is sufficiently large, shifting the strobe pulse may turn out to be useful in vibration analysis.

8.1.3 Phase Modulation of the Reference Wave

The possibilities inherent in the method of *Powell* and *Stetson* can be broadened by modulating the phase of the reference wave during hologram recording [8.25-29]. As will be shown below, by modulating the phase of the reference wave, one can both improve and reduce the sensitivity of the time-average method.

As we have already seen, in the time-average method the light wave scattered from a point Q of a harmonically vibrating object is phase-modulated and the depth of modulation is proportional to the vibration amplitude of this point. Just as before, we write the complex amplitude of the wave coming to the hologram from the vibrating object in the form

$$a_0 \exp\left\{-i\left[\phi_0 + \frac{2\pi}{\lambda}(\cos\psi_1 + \cos\psi_2) A \sin\omega t\right]\right\}. \tag{8.27}$$

If the reference wave is modulated in phase with the same frequency ω, for instance, by shifting the reference source, we find for its complex amplitude

$$a_r \exp\left\{-i\left[\theta_r + \frac{2\pi}{\lambda} M \sin\omega t\right]\right\}, \tag{8.28}$$

217

where M is the vibration amplitude of the reference source determining the phase modulation depth of the reference wave, and a_r and θ_r are the amplitude and initial phase of the reference wave.

The intensity at point Q in the reconstructed image is determined by the phase difference between object and reference waves:

$$I = \left| \frac{a_0}{\tau} \int_0^T \exp\left[-i\left\{ \frac{2\pi}{\lambda}[(\cos\psi_1 + \cos\psi_2)A - M]\sin\omega t + (\phi_0 - \theta_r) \right\} \right] dt \right|^2 .$$

$$(8.29)$$

Limiting ourselves to the case $\psi_1 = \psi_2 = 0$ and $\phi_0 = \theta_r$ we have

$$I = \left| \frac{a_0}{\tau} \int_0^T \exp\left\{ -i\left[\frac{4\pi}{\lambda}(A - \tfrac{1}{2}M)\sin\omega t \right] \right\} dt \right|^2 .$$

$$(8.30)$$

We see that (8.30) is similar to (8.4) (for the case $\psi_1 = \psi_2 = 0$), the only difference being that the amplitude A is replaced by A–M/2 in (8.30). Therefore, after some algebra similar to that applied to (8.6,7), we finally obtain

$$I = a_0^2 J_0^2 \left[\frac{4\pi}{\lambda}(A - \tfrac{1}{2}M) \right] .$$

$$(8.31)$$

Thus for the time-average method with a phase-modulated reference wave we have obtained the same relation (8.9), the only difference being that the argument of the zero-order Bessel function now depends on the phase modulation of the reference wave.

Consider the application of phase-modulated reference waves to the measurement of large vibration amplitudes. As seen from (8.31), by properly choosing the modulation depth of the reference wave, one can compensate the corresponding part of the object amplitude. Then the brightest fringes in the reconstructed image will be those with A–M/2 = 0 rather than with A = 0, as is the case with the method of *Powell* and *Stetson*. Each fringe becomes the locus of points of equal difference A–M/2.

This result was first demonstrated experimentally by *Aleksoff* [8.25] in a study of the vibrational mode of ADP crystals in which standing waves were excited. Phase modulation of the reference wave was used successfully by *Aleksoff* [8.26] to investigate loudspeaker vibrations with amplitudes in excess of 6λ. As shown by *Belgorodskii* et al. [8.27], reference-beam modulation permits the investigation of inhomogeneities in object vibrations containing a large translational component. They also analyzed surface vibrations of a piezoceramic electromechanic transducer with amplitudes exceeding 10λ.

Mottier [8.28] proposed a method of automatic modulation of the reference-wave phase by means of a small mirror fixed to the moving object. As shown by *Caulfield* [8.30], the method proposed by *Mottier* is a

218

particular case of "local reference beam holography" [8.31,32]. This term covers all methods involving the use of a reference beam formed by a part of the object beam. Any motion of the object illuminated by light from a laser is accompanied by a phase change in the optical field of the reference beam. In his experiment, *Mottier* obtained a hologram of a creeping snail, with the reference beam produced by a small mirror glued to its shell.

Mottier [8.29] furthermore proposed to perform synchronous phase modulation of the reference beam with a mirror mounted on a separate piezoelectric vibrator. The second piezoelectric vibrator was fixed to the cantilever beam to be studied. This method of reference-beam phase modulation was also used by *Belgorodskii* et al. [8.27].

Phase modulation of the reference wave can also be employed to measure small vibration amplitudes. If we choose the modulation depth M/2 in (8.31) such that the function $J_0^2(M/2)$ passes through the first root, one will be able to observe small amplitude regions as bright spots against a dark background of the object image.

The function $J_0^2[4\pi/\lambda\cdot(A-\frac{1}{2}M)$ in (8.31) can be expanded in a Taylor series for small A:

$$J_0\left[\frac{4\pi}{\lambda}(A-\tfrac{1}{2}M)\right] = J_0\left[\frac{4\pi}{\lambda}\frac{M}{2}\right] - \frac{4\pi}{\lambda}AJ_0'\left[\frac{4\pi}{\lambda}\frac{M}{2}\right]$$

$$+ \left(\frac{4\pi}{\lambda}\right)^2\frac{A^2}{2}J_0''\left[\frac{4\pi}{\lambda}\frac{M}{2}\right] - \dots . \qquad (8.32)$$

Limiting ourselves to the first two terms in (8.32) and taking into account that for the chosen value of M/2 the first term of the expansion vanishes, and $J_0'(z) = -J_1(z)$ we can write for the irradiance distribution in the reconstructed image:

$$I = a_0^2(4\pi A/\lambda)^2 J_1^2\left[\frac{4\pi}{\lambda}\frac{M}{2}\right] . \qquad (8.33)$$

Thus, it follows that the brightness distribution in the reconstructed image is proportional to the square of the amplitude A rather than to the square of the zero-order Bessel function. The "dark field" method permits measuring vibrations with an amplitude on the order of 0.01λ. Thus, in order to visualize very small amplitude vibrations, one can improve the sensitivity of the time-average method by phase-modulating the reference beam during the exposure. Implementation of this method in actual practice, however, requires a rather sophisticated setup.

In place of reference beam phase modulation, one can also measure small vibration amplitudes by means of a relatively simple "holographic subtraction" method proposed by *Gabor* et al. [8.33]. In this method, two holograms of the object are consecutively recorded on one photographic plate, with one of the beams shifted in phase by 180° between the exposures. The regions on the object which do not change between the exposures, vanish in reconstruction, so that one sees only the areas which have changed between the two exposures or during one of them [8.34].

To obtain a 180° phase shift in the object beam, *Wall* [8.35] proposed to superpose the hologram of the object in steady state onto the hologram of the vibrating object constructed by the method of *Powell* and *Stetson*. A similar technique, but intended for real-time observation of fringe patterns, was proposed by *Biedermann* and *Molin* [8.36]. *Hariharan* [8.37] achieved a 180° phase shift in the reference beam between two exposures by means of a half-wavelength plate.

8.1.4 Techniques of Holographic Vibration Analysis

Here we consider some technical aspects for studying vibrations by holographic interferometry, namely, the excitation of vibrations, the synchronization of laser pulses with object vibrations in the stroboscopic technique, and the reference wave modulation in the phase modulation method. The techniques of hologram recording, image reconstruction, and fringe interpretation do not differ markedly from those used in the investigation of fixed objects.

As already mentioned, the time-average method simply involves obtaining a hologram of the vibrating object in an exposure time many times longer than the oscillation period. It is essential that the frequency and amplitude of the oscillations be stabilized with sufficient accuracy. Forced oscillations of the object are excited, as a rule, by means of piezoelectric transducers, magnetostriction, or electromagnetic vibrators. The oscillations are transmitted to the object either directly (with the piezoelectric crystal glued onto the object on its unilluminated side), or with the aid of a special rod or needle. In this case the efficiency of power transmission from the vibrator to the object may depend on the point of contact. That is, if it occurs on the nodal line no oscillations can be excited. Therefore, the best option is probably the acoustic method for excitation by means of a sufficiently powerful electroacoustic transducer.

The key problem in a stroboscopic holographic study is the production of light pulses which are rigidly synchronized with the object vibrations. For this purpose one can use various mechanical choppers, however, this limits the frequency range to usually 10-20 kHz. A typical mechanical system of this type was described by *Ostrovsky* et al. [8.38]. The principal component is a disc with several rows of holes which chop both the laser beam used to record a hologram and the light beam supplied by an auxiliary light source. The latter produces a reference voltage which powers an electroacoustic transducer that excites oscillations in the object under study.

This system also permits real-time observation of interference fringes. In contrast to the method of *Powell* and *Stetson* [8.18], no fringe contrast degradation occurs in this case. For real-time observation, a hologram is recorded of the object at rest with the disc stopped, after which the disc is set in rotation and the vibrating object is viewed through the in-situ developed hologram. Thus, by gradually increasing the disc rotation one can follow the

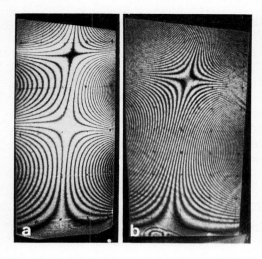

Fig.8.10. Interferogram of axial-flow compressor blades obtained by stroboscopic holography at (a) 4730 Hz, and (b) 5250 Hz (Courtesy of Laboratory for Strength of Materials, Nevskii Zavod Association, Leningrad)

dynamic development of the vibrations as the frequency of the driving force is varied. A disc-type stroboscope was employed in vibration studies of circular membranes and a cantilever beam [8.39, 40]. Its major drawback, just as in other mechanical arrangements, is the difficulty of adjusting the desired disc rotation frequency and its stabilization, in addition to the limited available frequency range and the impossibility of continuously varying the duration of the strobe pulse.

A more appropriate method makes use of a stroboscope with a Kerr-cell modulator [8.4, 22]. In devices of this type the frequencies of the object vibration and of the strobe pulses are controlled by an acoustic generator which excites object vibrations through a piezo- or electromechanical vibrator, and also drives a delayed square-pulse generator to operate the Kerr cell. The strobe pulse can be shifted relative to the object oscillation phase. Figure 8.10 presents two interferograms obtained by stroboscopic holography.

A pulsed laser which can be synchronized with any phase of the vibration of the object under study can also be employed to obtain interferograms similar to those recorded in stroboscropic holography [8.41]. In this case, the fringe contrast is not affected by instabilities in the oscillation frequency and amplitude, since hologram recording occurs only during a small part of the oscillation period.

The reference wave can be phase-modulated by means of a small lightweight mirror mounted on the vibrator oscillating at the frequency of the object under investigation. If this mirror is mounted on the vibrating object itself and oscillates with it, then, as already mentioned, we have the setup for local reference-beam holography. In this method, while rigid translational as well as oscillatory displacements of the object are totally excluded, the possibility of adjusting the phase shift and modulation depth in order to improve or reduce sensitivity, is lost.

8.2 Vibration Studies of Rotating Objects

A specific case of application of holographic vibration analysis is the study of rotating objects. This field is directly applicable to the industrial environment. Because of the various difficulties involved in such studies, particularly associated with the necessity to completely compensate for the rotation of the object as a whole, the object is usually removed from its normal operating environment and mounted in a special stand which, while allowing vibrational excitation, does not respond to the environment. Some methods of studying rotating objects have been proposed and will be discussed in the next section.

8.2.1 Reducing the Exposure Time

The most radical and general method for holographing rotating (or, more generally speaking, moving) objects is to reduce the exposure time, e.g., by using pulsed lasers as light sources. A straightforward calculation based on the requirement that the change of the optical path from the object to the hologram during an exposure be not greater than $\lambda/4$ suggests that for a pulse duration of 20 ns, the object's velocity component directed at the hologram should not exceed 10 m/s.

These requirements can be weakened when the object motion can be compensated by means of the above-mentioned local reference-beam method [8.42]. In this method, the reference wave is formed by reflection from a mirror mounted on the moving object. Thus, the phase change during the exposure is measured from the point on the object where the mirror is mounted. This technique was used by *Waters* [8.43] who produced the reference wave by focussing the light onto a small spot on the rotating object, serving as the "local reference source". In order to obtain a holographic interferogram, however, one must make two exposures corresponding to the same position of both the vibrating and fixed objects. Therein lies the major difficulty in constructing holographic interferograms of rotating objects. Figure 8.11 displays an interferogram of a rotating model for a fan wheel (rotation speed 1500rpm) vibrating at 531 Hz [8.44].

Fig.8.11. Vibration of a rotating model of a ventilator wheel. Frequency 531 Hz. Rotation velocity 1500 rpm [8.44]

Kawase et al. [8.45] described a method to ensure the positioning of the specimen to exactly the same position that it occupied during the rotation at the moment of the first exposure. The speckle pattern formed by a diffuse transparency which is fixed rigidly to the object is recorded simultaneously with the first exposure during the object rotation. The object is stopped and the speckle photograph developed in situ and illuminated through the diffuse transparency with a parallel beam from a He-Ne laser. One observes an interference-ring pattern in the focal plane of a lens placed behind the speckle photograph in the path of the laser beam. The transparency and the object can be returned into their initial positions by matching the center of these rings with the lens focus. The second exposure of the hologram is made in this position.

Another technique to match the object positions during the two exposures is to use a pulsed laser synchronized with the position of the object. Various synchronizing transducers which control the operation of the laser Q-switch [8.46, 47] can be used. If the Q-switch is based on a rotating prism, synchronization is automatically achieved if the prism is mounted directly on the object shaft.

Bjelkhagen and *Abramson* [8.48] used a sandwich hologram made up of two photographic plates mounted on the rotation shaft for matching the object positions in two exposures. Matching is achieved by rotating one plate with respect to the other. The same result can be achieved with a two-exposure hologram recorded on a single photographic plate. This requires, in the image reconstruction step, a special rotational shift interferometer (Fig.8.12) which separates the reconstructed image into two identical channels and rotates one of them (for instance, by means of a Dauvais prism) until an exact matching of the images obtained in the two exposures has been achieved. The match is observed when interference fringes appear.

8.2.2 The Derotator

Holographic interferograms of rotating objects can also be obtained by using a so-called derotator which compensates for the rotation of the object and forms its fixed image [8.49-51]. In the simplest case, the derotator is an erecting Dauvais prism similar to the one shown in Fig.8.12 which rotates with half the angular velocity of the object rotation. An arrangement in-

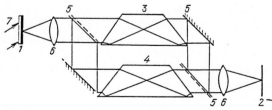

Fig.8.12. Rotational shift interferometer: *1* - two-exposure image hologram of a rotating object, *2* - interferogram recording plane, *3* - fixed Dauvais prism, *4* - Dauvais prism capable of rotating around the optical axis, *5* - Mach-Zehnder interferometer mirrors, *6* - objective lenses, *7* - reconstructing beam

223

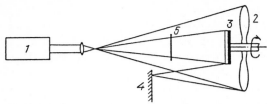

Fig.8.13. Recording the interferogram of a rotating object on a photographic plate mounted in the Leith-Upatnieks off-axis configuration: *1* - pulsed laser, *2* - rotating object, *3* - photographic plate, *4* - mirror, *5* - filter for correction for reference beam intensity

Fig.8.14. Holography of a rotating object on a photographic plate mounted in the opposed-beam configuration of Denisyuk

cluding a derotator and a pulsed ruby laser has been used to record two-exposure holograms of vibrating objects rotating at frequencies up to 150 Hz. *McBain* et al. [8.50] used a derotator and a high-power argon laser for real-time holography of rotating objects.

The derotator prism must be fabricated with a high degree of precision to prevent the appearance of spurious fringes. Considerable difficulties are associated with the necessity of very precise synchronization between the angular velocities of the object and the derotator prism.

8.2.3 Rotating Hologram

Another method for holographic interferometry of rotating objects is to fix the photographic plate at the center of the rotating object which thus rotates with the latter [8.48, 52]. The setup is shown schematically in Fig. 8.13. Holograms were obtained in the Leith-Upatnieks off-axis reference-beam configuration with a pulsed Q-switched ruby laser. This setup has the obvious drawback that only half of the object can be holographed at a time.

A simpler way is to use Denisyuk's arrangement (Fig.8.14) [8.53]. Here, no mirror is required to generate the object beam, the photographic plate is fixed directly to the object, and the hologram reconstructs the whole image of the object. It has been shown [8.53] that the accuracy $\Delta\alpha$ with which the axis of object and hologram rotation and that of the illuminating beam can be matched should satisfy the condition

$$\Delta\alpha < \lambda[4d\tan\beta\sin(\Delta\psi/2)]^{-1} , \tag{8.34}$$

where d is the distance from the object to the photographic plate, β is the maximum angle of light scattering (reflecting) from the diffuse surface of

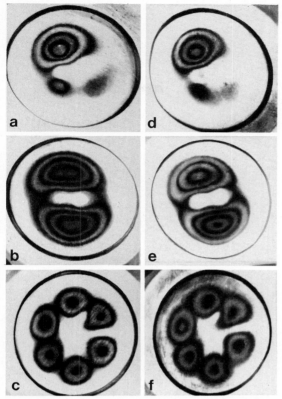

Fig.8.15. Interferogram of a vibrating membrane recorded with LG-38 laser in CW mode. (**a, b, c**) the membrane is rotating; (**d, e, f**) the membrane is fixed. Excitation frequency: (**a, d**) 3050 Hz, (**b, e**) 5950 Hz, (**c, f**) 10500 Hz; rotation frequency 100 rad/s

the object, and $\Delta\psi$ is the angle through which the object and the recording medium rotate during the hologram recording. Upon inspection of this relation, the following conclusions can be drawn:

1) The photographic plate should be positioned as close to the object as possible;

2) to make the angle β smaller, retroreflective paints having a narrow reflection indicatrix should be used;

3) The angle $\Delta\psi$ through which the object turns during the exposure should be made as small as possible.

Morozov et al. [8.53] employed the arrangement of Fig.8.14 to obtain holographic interferograms of a vibrating membrane which, at the same time, rotates with a velocity of up to 100 rad/s (Fig.8.15), with a cw HeNe-laser. No difference in quality between the interferograms of a rotating and a fixed membrane was noticed.

References

Chapter 1

1.1   D. Gabor: A new microscopic principle. Nature **161**, 777-778 (1948)
1.2   E. Leith, J. Upatnieks: Wavefront reconstruction with continuous-tone objects. J. Opt. Soc. Am. **53**, 1377-1381 (1963)
1.3   E. Leith, J. Upatnieks: Wavefront reconstruction with diffused illumination and three-dimensional objects. J. Opt. Soc. Am. **54**, 1295-1301 (1964)
1.4   E. Leith, J. Upatnieks: New technique in wavefront reconstruction. J. Opt. Soc. Am. **51**, 1469 (1961)
1.5   E. Leith, J. Upatnieks: Reconstructed wavefronts and communication theory. J. Opt. Soc. Am. **52**, 1123-1130 (1962)
1.6   Yu.N. Denisyuk: On the reproduction of the properties of an object in the wavefield of the radiation scattered by it. Dokl. AN SSSR **144**, 1275-1276 (1962)
1.7   Yu.N. Denisyuk: On the reproduction of the optical properties of an object by the wavefield of its scattered radiation. Opt. Spektrosk. **15**, 522-532 (1963)
1.8   J.-C. Viénot, P. Smigielski, H. Royer: *Holographie optique* (Dunod, Paris 1971)
1.9   R.J. Collier, C.B. Burckhardt, L.H. Lin: *Optical Holography* (Academic, New York 1971)
1.10 Y.I. Ostrovsky, M.M. Butusov, G.V. Ostrovskaya: *Interferometry by Holography*, Springer Ser. Opt. Sci., Vol.20 (Springer, Berlin, Heidelberg 1980)
1.11 M. Françon: *Holographie* (Masson, Paris 1969)
1.12 J.N. Butters: *Holography and its Technology* (Peregrinus, London 1971)
1.13 V.M. Ginzburg, B.M. Stepanov (eds.): *Golografiya: Metody i Apparatura* (holography: Methods and equipment) (Sov. Radio, Moscow 1974)
1.14 Y.I. Ostrovsky: *Holography and its Applications* (Mir, Leningrad 1977)
1.15 C.M. Vest: *Holographic Interferometry* (Wiley, New York 1979)
1.16 W. Schumann, M. Dubas: *Holographic Interferometry from the Scope of Deformation Analysis of Opaque Bodies*. Springer Ser. Opt. Sci., Vol.16 (Springer, Berlin, Heidelberg 1979)
1.17 R.J. Jones, C. Wykes: *Holographic and Speckle Interferometry* (Cambridge Univ. Press, Cambridge 1983)
      J.C. Dainty (ed.): *Laser Speckle and Related Phenomena*, 2nd edn., Topics Appl. Phys., Vol.9 (Springer, Berlin, Heidelberg 1984)
1.18 A.G. Kozachok: *Golograficheskiye metody issledovaniya v eksperimentalnoi mekhanike* (holographic methods of research in experimental mechanics) (Mashinostroyenie, Moscow 1984)
1.19 I.S. Klimenko: *Golografiya sfokusirovannykh izobrazhenii i spekl-interferometriya* (image holography and speckle interferometry) (Nauka, Moscow 1985)
1.20 W. Schumann, J.-P.Zürcher, D. Cuche: *Holography and Deformation Analysis* (Springer, Berlin, Heidelberg, 1985)
1.21 Ju.I. Ostrowski: *Holografie-Grundlagen, Experimente und Anwendungen* (Deutsch, Frankfurt/Main 1988)
1.22 Sh.D. Kakichashvili: On the polarization recording of holograms. Opt. Spektrosk. **33**, 324-327 (1972)

1.23 A.K. Rebane, R.K. Kaarti, P.M. Saap: Dynamic picosecond-range holography by photochemical hole burning. Pisma Zh. Tekh. Fiz. **38**, 320-323 (1983)

1.24 R.R. Milier: Cardinal points and the novel imaging properties of a holographic system. J. Opt. Soc. Am. **56**, 219-223 (1966)

1.25 O. Bryngdahl, A. Lohmann: Nonlinear effects in holography. J. Opt. Soc. Am. **58**, 1325-1334 (1968)

1.26 B.G. Turukhano, N. Turukhano: Interferometric testing of holographic equipment. Zh. Tekh. Fiz. **38**, 757-758 (1968)

1.27 J.E. Sollid, I.B. Swint: A determination of the optimum beam ratio to produce maximum contrast photographic reconstruction from double-exposure holographic interferograms. Appl. Opt. **9**, 2717-2719 (1970)

1.28 I.F. Budagyan, V.F. Dubrovin, S.N. Kanalyuk, R.I. Mirovitskii, V.V. Usatyuk: Holographic refractometer and reflectometer. Priboryi Tekh. Eksp. **6**, 174-177 (1972)

1.29 Yu.I. Ostrovskii: *Golografiya (Holography)* (Nauka, Leningrad 1970)

1.30 J.A. Gilbert, T.D. Dudderur, M.E. Schultz, A.J. Boehnlin: The monomode fiber - a new tool for holographic interferometry. Exp. Mech. **23**, 190-195 (1983)

1.31 J.A. Gilbert, I.W. Herrick: Holographic displacement analysis with multimode fiber optics. Exp. Mech. **21**, 315-320 (1981)

1.32 J.A. Gilbert, M.E. Schultz, A.I. Boehnlin: Remote displacement analysis using multimode fiber-optic bundles. Exp. Mech. **22**, 398-400 (1982)

1.33 T.R. Hsu, R.G. Moyer: Application of fiber optics in holography. Appl. Opt. **10**, 669-670 (1971)

1.34 A.M.P.P. Leite: Optical fibre illuminators for holography. Opt. Commun. **28**, 303-308 (1979)

1.35 M. Takahashi, S. Sugiymi: Toyota tries fiber holography. Laser Focus 7, 29 (May 1971)

1.36 M. Yonemura, T. Mishisaka, H. Machid: Endosconic hologram interferometry using fiber optics. Appl. Opt. **20**, 1664-1667 (1981)

1.37 N. Abramson: *The Making and Evaluation of Holograms* (Academic, London 1981)

1.38 N.I. Borisenok, M.M. Ermolaev, Yu.A. Lyapin et al.: External Factors in Bulk Reflection Hologram Recording, in *Opticheskaya Golografiya: Pratkicheskiye Primineniya* (optical holography: Applications), ed. by Yu.N. Denisyuk (Nauka, Leningrad 1985) pp.41-50

1.39 R.R. Gerke, Yu.N. Denisyuk, V.A. Kozak, V.I. Lokshin, V.M. Tikhonov: SIN interferometric bench for hologram recording and study. Opt. Mekh. Prom. **8**, 70-71 (1971)

1.40 V.M. Ginzburg, B.M. Stepanov: *Golograficheskiye Izmereniya* (holographic measurements) (Sov. Radio, Moscow 1982)

1.41 R.F. Erf (ed.): *Holographic Nondestructive Testing* (Academic, New York 1974)

1.42 H.J. Caulfield (ed.): *Handbook of Optical Holography* (Academic, New York 1979)

1.43 C.M. Vest: Holographic interferometry: Some recent developments. Opt. Eng. **19**, 654-658 (1989)

1.44 I.V. Volkov: A study of the deformation of full-scale specimens in stress concentration areas by holography. Probl. Prochn. **12**, 92-95 (1974)

1.45 V.G. Seleznev: Holographic attachment to a machine for displacement measurements. Zavod. Labor. **46**, 764-766 (1980)

1.46 A.K. Baev, V.E. Bortsov, A.A. Kapustin: A study of the strength properties of cutting tools by holography, in *Prikladnye Voprosy Golografii* (holographic applications) (LINP, Leningrad 1982) pp.221-233

1.47 S.B. Artemenko, V.P. Ushakov, A.N. Chernovol: All-purpose holographic camera. Zavod. Labor. **49**, 88-89 (1983)

1.48 D.M. Rowley: A holographic interference camera. J. Phys. E **12**, 971-975 (1979)

1.49 N.M. Ganzherli, S.B. Gurevich, V.V. Kovalenok et al.: Holographic study of processes and objects on the manned spacecraft Salyut-6. Zhur. Tekh. Fiz. **52**, 2192-2197 (1982)

1.50 M. Born, E. Wolf: *Principles of Optics*, 5th edn. (Pergamon, Oxford 1975)

1.51 N. Abramson: The "Holo-diagram": A practical device for making and evaluation of holograms, in *The Engineering Uses of Holography*, ed. by E.R. Robertson, J.M. Harvey (Cambridge Univ. Press, Cambridge 1970) pp.45-46

1.52 W. Koechner: *Solid-State Laser Engineering*, 2nd edn., Springer Ser. Opt. Sci., Vol.1 (Springer, Berlin, Heidelberg 1988)

1.53 V.D. Petrov: Express treatment of photographic layers in reflection hologram production. Zhur.nauchn. i prikl. fotogr. i kinematogr. **21**, 214 (1976)

1.54 P. Hariharan: Holographic recording materials: Recent development. Opt. Eng. **19**, 636-641 (1980)

1.55 A.A. Frisem, Y. Katziz, Z. Rav-Noy, B. Sharon: Photoconductor-thermoplastic devices for holographic nondestructive testing. Opt. Eng. **19**, 659-665 (1980)

1.56 M.P. Petrov, S.I. Stepanov, A.V. Khomenko:

Chapter 2

2.1 C.M. Vest: *Holographic Interferometry* (Wiley, New York 1979)

2.2 W. Schumann, M. Dubas: *Holographic Interferometry from the Scope of Deformation Analysis of Opaque Bodies*, Springer Ser. Opt. Sci., Vol.16 (Springer, Berlin, Heidelberg 1979)

2.3 M.H. Hormann: An application of wavefront reconstruction to interferometry. Appl. Opt. **4**, 333-336 (1965)

2.4 L.H. Tanner: A study of fringe clarity in laser interferometry and holography. J. Phys. E **1**, 517-522 (1968)

2.5 W.C. Gates: Holographic phase recording by interference between reconstructed wavefronts from separate holograms. Nature **220**, 473-474 (1968)

2.6 A. Havener, R. Radley: Dual hologram interferometry. Opt. Electron. **4**, 48-52 (1972)

2.7 N. Abramson: Sandwich hologram interferometry: A new dimension in holographic comparison. Appl. Opt. **13**, 2019-2025 (1974)

2.8 N. Abramson: Sandwich hologram interferometry: 2. Some practical calculations. Appl. Opt. **14**, 981-984 (1975)

2.9 N. Abramson: Sandwich hologram interferometry. 4. Holographic studies of two milling machines. Appl. Opt. **16**, 2521-2531 (1977)

2.10 K.A. Stetson, R.L. Powell: Hologram interferometry. J. Opt. Soc. Am. **56**, 1161-1166 (1966)

2.11 M.M. Butusov, Yu.G. Yurkevich: A simple arrangement for real-time holographic interferometry. Zh. Nauch. Prikl. Fotogr. Kinematogr. **4**, 303-305 (1971)

2.12 Y.I. Ostrovsky, M.M. Butusov, G.V. Ostrovskaya: *Interferometry by Holography*, Springer Ser. Opt. Sci., Vol.20 (Springer, Berlin, Heidelberg 1980)

2.13 B.U. Achia: A simple plate holder for on site wet processing of holograms in real-time holographic interferometry. J. Phys. E **5**, 128-129 (1972)

2.14 W. Van Deelen, P. Nisenson: Mirror blank testing by real-time holographic interferometry. Appl. Opt. **8**, 951-955 (1969)

2.15 N.G. Vlasov, K.N. Petrov, V.A. Marinovskii, A.E. Shtan'ko: On off-bench holographic interferometry, in *Optiko-Geometricheskiye Metody Issledovaniya Deformatsii i Napryazhenii* (optic-geometrical methods of strain and stress analysis) (DGU, Dnepropetrovsk 1978) pp.54-56

2.16 N. Maclead, D.N. Kapur: A kinematically designed mount for the precise location of specimens for holographic interferometry. J. Phys. E **6**, 423-424 (1973)

2.17 I.E. Furse: Kinematic design of fine mechanism in instruments. J. Phys. E **14**, 264-271 (1981)

2.18 K. Biedermann, N.-E. Molin: Combining hypersensitization in situ processing for time-average observation in real-time hologram interferometry. J. Phys. E **3**, 669-681 (1970)

2.19 A.A. Friesem, I.L. Walker: Experimental investigation of some anomalies in photographic plates. Appl. Opt. **8**, 1504-1506 (1969)

2.20 A.B. Kudrin, P.I. Polukhin, N.A. Chichenev: *Golografiya i Deformatsiya Metallov* (holography and deformation of metals) (Metallurgiya, Moscow 1982)

2.21 Yu.M. Cherkasov, Yu.A. Pryakhin: *Photothermoplastic Processes for Holography Based on High-Resolution, Semitransparent Films*, ed. by G.A. Sobolev (Nauka, Leningrad 1979) pp.119-143

2.22 B. Ineichen, C. Liegeois, P. Meyrueis: Thermoplastic film camera for holography recording for extended objects in industrial application. Appl. Opt. **21**, 2209-2214 (1982)

2.23 K. Iwata, R. Nagata: Fringe formation in multiple-exposure holographic interferometry. Appl. Opt. **26**, 995-1007 (1979)

2.24 J.M. Burch, A.E. Ennos, R.J. Wilton: Dual and multiple beam interferometry by wavefront reconstruction. Nature, **209**, 1015-1016 (1966)

2.25 K.A. Stetson: Fringe interpretation for hologram interferometry of rigid-body motion and homogeneous deformation. J. Opt. Soc. Am. **64**, 1-10 (1974)

2.26 J.D. Redman: Holographic velocity measurement. J. Sci. Instrum. **44**, 1032- 1033 (1967)

2.27 R.L. Powell, K.A. Stetson: Interferometric analysis by wavefront reconstruction. J. Opt. Soc. Am. **55**, 1593-1598 (1965)

2.28 K.A. Stetson, R.L. Powell: Interferometric hologram evaluation and real-time vibration analysis of diffuse objects. J. Opt. Soc. Am. **55**, 1694-1695 (1965)

2.29 E.B. Aleksandrov, A.M. Bonch-Bruevich: Holographic study of surface deformation. Zhur. Tekh. Fiz. **57**, 360-369 (1967)

2.30 A.E. Ennos: Measurement of in-plane surface strain by hologram interferometry. J. Phys. E **1**, 731-734 (1968)

2.31 J.E. Sollid: Holographic interferometry applied to measurements of small static displacements of diffusely reflecting surfaces. Appl. Opt. **8**, 1587-1595 (1969)

2.32 N.L. Hecht, J.E. Minardi, D. Lewis: Quantitative theory for predicting fring pattern formation in holographic interferometry. Appl. Opt. **12**, 2665-2676 (1973)

2.33 N.-E. Molin, K.A. Stetson: Measurement of fringe loci and localization in hologram interferometry for pivot motion, in-plane rotation and in-plane translation. Pt.I. Optik, **31**, 157-177 (1970)

2.34 N.-E. Molin, K.A. Stetson: Measurement of fringe loci and localization in hologram interferometry for pivot motion, in-plane rotation and in-plane translation. Pt.II. Optik **31**, 281-291 (1970)

2.35 I. Prikryl: Localization of interference fringes in holographic interferometry. Opt. Acta **21**, 675-681 (1974)

2.36 T. Tsuruta, N. Shiotake, Y. Itoh: Formation and localization of holographically produced interference fringes. Opt. Acta **16**, 723-733 (1969)

2.37 S. Walles: Visibility and localization of fringes in holographic interferometry of diffusely reflecting surfaces. Ark. Fys. **40**, 299-403 (1970)

2.38 M. Yonemura: Geometrical theory of fringe localization in holographic interferometry. Opt. Acta **27**, 1537-1549 (1980)

2.39 K.A. Stetson: A rigorous treatment of the fringes of hologram interferometry. Optik **4**, 386-400 (1969)

2.40 R.F. Erf (ed.): *Holographic Nondestructive Testing* (Academic, New York 1974)

2.41 F. Grünewald, D. Kaletsch, V. Lehmann: Holographische interferometrie und deren quantitative Auswertung, demonstriert am Beispiel zylindrisher GEK-Rohre. Optik 37, 102-110 (1973)

2.42 D.L. Mader: Holographic interferometry on pipes: precision interpretation by least-squares fitting. Appl. Opt. 24, 3784-3790 (1985)

2.43 I.A. Birger, Ya.G. Panovko (eds.): *Prochnost, Ustoichivost, Kolebaniya*, T.1 (strength, stability vibrations, Vol.1) (Mashinostroyeniye, Moscow 1968)

2.44 P.M. Boone: Use of reflection holograms in holographic interferometry and speckle correlation for measurement of surface displacement. Opt. Act. 22, 579-589 (1975)

2.45 D.B. Neumann, R.O. Penn: Off-table holography. Exp. Mech. 25, 241-244 (1975)

2.46 A.E. Shtan'ko, M.N. Guzikov: Compact holographic interferometer for residual stress evaluation in structures, in *Ostatochnye Tekhnologicheskiye Napryazheniya* (residual technological stress) (AN SSSR, Moscow 1985) pp.366-370

2.47 V.J. Corcoran, R.W. Herron, J.G. Jaramillo: Generation of a hologram from a moving target. Appl. Opt. 5, 668-669 (1966)

2.48 J.P. Waters: Object motion compensation by speckle reference beam holography. Appl. Opt. 11, 630-636 (1972)

2.49 E. Champagne, L. Kersch: Control of holographic interferometric fringe patterns. J. Opt. Soc. Am. 59, 1535 (1969)

2.50 L.A. Kersch: Advanced concepts of holographic nondestructive testing. Mater. Eval. 29, 125-129 (1971)

2.51 A. Stimpfling, P. Smigielski: New method for compensating and measuring any motion of three-dimensional objects in holographic interferometry. Opt. Eng. 24, 821-823 (1985)

2.52 N. Abramson, H. Bjelkhagen: Sandwich hologram interferometry: 5. Measurement of in-plane displacement and compensation for rigid body motion. Appl. Opt. 18, 2870-2880 (1979)

2.53 I.S. Klimenko, V.P. Ryabukho: Spatial filtration and holographic interferometry, in *Voprosy Prikladnoi Golografii* (problems in applied holography) (LNIP, Leningrad 1982) pp.62-80

2.54 I.S. Klimenko, V.P. Ryabukho, B.V. Feduleev: On the separation of information concerning different motions in holographic interferometry based on spatial filtration. Opt. Spektr. 55, 140-147 (1983)

2.55 J.A. Gilbert, D.T. Vedder: Development of holographic techniques to study thermally induced deformation. Exp. Mech. 21, 138-144 (1981)

2.56 L.O. Heflinger, R.F. Wuerker, H. Spetzler: Thermal expansion coefficient measurement of diffusely reflecting, samples by holographic interferometry. Rev. Sci. Instrum. 44, 629-633 (1973)

2.57 S.A. Novikov, V.P. Shchepinov, V.S. Aistov: Study of residual deformation development in a MBNLZ roller band by coherent optics, in *Dinamikai Prochnost Metallurgicheskikh Mashin* (dynamics and strength of metallurgical equipment) (VNIIMETMASH, Moscow 1984) pp.47-52

2.58 T.R. Hsu, R.G. Moyer: Application of holography in high temperature displacement measurements. Exp. Mech. 12, 431-432 (1972)

2.59 K.A. Heines, B.P. Hildebrand: Surface-deformation measurement using the wavefront reconstruction technique. Appl. Opt. 5, 595-602 (1966)

2.60 C.H.F. Velzel: Fringe contrast and fringe localization in holographic interferometry. J. Opt. Soc. Am. 60, 419-420 (1970)

2.61 A.A. Voevodin, V.P. Kazak, I.N. Nagibina: Interpretation of holographic interferograms in surface strain measurements. Zhur. Tekh. Fiz. 52, 703-708 (1982)

2.62 V.A. Zhilkin: Optical interference methods in strain studies. Zavod. Labor. 10, 57-63 (1981)

2.63 P. Hariharan, B.F. Oreb, N. Brown: Real-time holographic interferometry: A microcomputer system for the measurement of vector displacements. Appl. Opt. 22, 876-880 (1983)

2.64 S. Millmore, J.A. Allsop: A qualitative investigation of holographic interferometry techniques applied to the measurement of general displacement field. Strain 14, 106-111 (1978)

2.65 S.K. Dhir, J.P. Sikora: An improved method for obtaining the general displacement field from a holographic interferogram. Exp. Mech. 12, 323-327 (1972)

2.66 Yu.I. Ostrovsky: Golografiya (holography) (Nauka, Leningrad 1970)

2.67 L.A. Borynyak, S.I. Gerasimov, V.A. Zhilkin: Experimental techniques of interferogram recording and interpretation providing the required accuracy in strain tensor component determination. Avtometriya 1, 17-24 (1982)

2.68 A.E. Ennos, M.S. Virdee: Application of reflection holography to deformation measurement problems. Exp. Mech. 22, 202-209 (1982)

2.69 P. Wesolowski: Some aspects of numberical in-plane strain analysis by reflection holography. Optik 71, 113-118 (1985)

2.70 Z. Füzessy: Improvement of the holographic method of displacement field measurement. Zhur. Tekh. Fiz. 49, 399-403 (1979)

2.71 Y.Y. Hung, C.P. Hu, D.R. Menley: Two improved methods of surface-displacement measurements by holographic interferometry. Opt. Commun. 8, 48-51 (1973)

2.72 K. Scibayama, H. Uchiyama: Measurement of three-dimensional displacements by hologram interferometry. Appl. Opt. 10, 2150-2154 (1971)

2.73 T. Matsumoto, K. Iwata, R. Nagata: Distortionless recording in double-exposure holographic interferometry. Appl. Opt. 12, 1660-1662 (1973)

2.74 R.J. Pryputniewicz: Determination of the sensitivity vectors directly from holograms. J. Opt. Soc. Am. 67, 1351-1353 (1977)

2.75 R. Pawluczyk, Z. Kraska: Diffuse illumination in holographic double-aperture interferometry. Appl. Opt. 24, 3072-3078 (1985)

2.76 M.R. Wall: Zero motion fringe identification, in Applications de l'Holographie, Proc. Symp. Besançon 1970, ed. by J.C. Viénot, J. Bulabois, J. Pasteur. Paper 4.9

2.77 U. Köpf: Fringe order determination and zero motion fringe identification in holographic displacement measurements. Opt. Laser Technol. 5, 111-113 (1973)

2.78 N. Abramson: The holo-diagram V: A device for practical interpreting of hologram interference fringes. Appl. Opt. 11, 1143-1147 (1972)

2.79 N.G. Vlasov, A.E. Shtan'ko: Fringe order and sign determination. Zhur. Tekh. Fiz. 46, 196-197 (1976)

2.80 H. Kohler: Holografische Interferometrie: V. Zur Problematik der Ordnungsbestimmung. Optik 60, 411-425 (1982)

2.81 I. Bogar: Calculation method to eliminate the sign problem of fringe order in holographic interferometry. Period. Polytechn. Mech. Eng. 24, 279-284 (1980)

2.82 B. Harnisch, W. Schreiber, L. Wenke: A method to avoid sign ambiguity in holographic interferometric displacement measurements. Opt. Acta 30, 699-703 (1983)

2.83 Yu.A. Rakushin, Yu.N. Solodkin: Interpretation of holographic interferograms, in Golograficheskiye Izmeritelnye Sistemy (holographic measurement systems), ed. by A. Kozachok, (NETI, Novosibirsk 1976) pp.41-47

2.84 R.J. Jones, C. Wykes: Holographic and Speckle Interferometry (Cambridge Univ. Press, Cambridge 1983)

2.85 I.M. Nagibina, V.V. Khopov: Automated processing of holographic interferograms for determination of diffusely reflecting surface displacement vector. Izv. Vuzov. Priborostroyeniye 26, 80-84 (1983)

2.86 R. Dändliker, B. Ineichen, F.M. Mottier: High resolution hologram interferometry by electronic phase measurement. Opt. Commun. 9, 412-416 (1973)

2.87 R. Dändliker: Heterodyne holographic interferometry. Progress in Optics 17 (North-Holland, Amsterdam 1980)

2.88 V.F. Bellani, A. Sona: Measurement of three-dimensional displacements by scanning a double exposure hologram. Appl. Opt. **13**, 1337-1341 (1974)

2.89 L. Ek, K. Biedermann: Analysis of a system for hologram interferometry with a continuously scanning reconstruction beam. Appl. Opt. **16**, 2535-2542 (1977)

2.90 L. Ek, K. Biedermann: Implementation of hologram interferometry with a continuously scanning reconstruction beam. Appl. Opt. **17**, 1727-1732 (1978)

2.91 A.E. Ennos, D.W. Robinson, D.C. Williams: Automatic fringe analysis in holographic interferometry. Opt. Acta. **32**, 135-145 (1985)

2.92 H. Kokler: Untersuchungen zur quantitativen Analyse der holographischen Interferometrie. Optik **39**, 229-235 (1974)

2.93 W. Osten: Some consideration on the statistical error analysis in holographic interferometry with application to an optimized interferometry. Opt. Acta. **32**, 827-838 (1985)

2.94 H. Kohler: Ein neues Verfahren zur quantitativen Auswertung holographischer Interferogrammen. Optik **47**, 135-152 (1977)

2.95 H. Kohler: General formulation of the holographic-interferometric evaluation methods. Optik **47**, 469-475 (1977)

2.96 H. Kohler: Interferenzliniedynamik bei der quantitativen Auswertung holographischer Interferogrammen. Optik **47**, 9-24 (1977)

2.97 H. Kohler: Zur Vermessung holographisch-interferometrischer Verformungsfelder. Optik **47**, 271-282 (1977)

2.98 N.G. Vlasov, O.G. Lisin, Yu.I. Savilova: A simplified method of calculating holographic interferograms of diffusely reflecting objects. Zhur. Tekh. Fiz. **54**, 2037-2039 (1984)

2.99 V.A. Zhilkin, L.A. Borynyak: Optical methods of determining small displacements and deformations in structural elements, in *Golograficheskiye Izmeritelnye Sistemy* (holographic measurement systems), ed. by A. Kozachok (NETI, Novosibirsk 1976) pp.76-92

2.100 P.W. King III: Holographic interferometry technique utilizing two plates and relative fringe orders for measuring microdisplacements. Appl. Opt. **13**, 231-233 (1974)

2.101 N.G. Vlasov, A.E. Shtan'ko: Problems in holographic interferometry, in *Proc. VIII All-Union School on Holography* (LINP, Leningrad 1976) pp.202-231

2.102 Z. Füzessy, P. Wesolowski: Simplified static holographic evaluation method omitting the zero-order fringe. Opt. Eng. **24**, 1023-1025 (1985)

2.103 D. Bijl, R. Jones: A new theory for the practical interpretation of holographic interference patterns resulting from static surface displacements. Opt. Acta **21**, 105-118 (1974)

2.104 R. Jones: An experimental verification of a new theory for the interpretation of holographic interference patterns resulting from static surface displacements. Opt. Acta **21**, 257-266 (1974)

2.105 W.C. Gate: Holographic measurement of surface distortion in three dimensions. Opt. Technol. **1**, 247-250 (1969)

2.106 P.M. Boone: Determination of three orthogonal displacement components from one double-exposure hologram. Opt. Laser Technol. **4**, 162-167 (1972)

2.107 P.M. Boone, L.C. De Backer: Determination of three orthogonal displacement components from one double-exposure hologram. Optik **37**, 61-81 (1973)

2.108 G.P. Monakhov-Ilyin, V.P. Shchepinov, B.A. Morozov, V.S. Aistov: Analysis of combined operation of structural elements by the finite element techniques and holographic interferometry. Vestnik Mashinostr. **6**, 9-11 (1980)

Chapter 3

3.1 S.K. Dhir, J.P. Sikora: An improved method for obtaining the general displacement field from a holographic interferogram. Exp. Mech. **12**, 323-327 (1972)

3.2 P.W. King III: Holographic interferometry technique utilizing two plates and relative fringe orders for measuring microdisplacements. Appl. Opt. **13**, 231-233 (1974)

3.3 H. Kohler: Untersuchungen zur quantitativen Analyse der holographischen Interferometrie. Optik **39**, 229-235 (1974)

3.4 C.A. Sciammarella, T.Y. Chiang: Holographic interferometry applied to the solution of shell problem. Exp. Mech. **14**, 217-224 (1974)

3.5 C.A. Sciammarella, J.A. Gilbert: Strain analysis of a disc subjected to diametric compression by means of holographic interferometry. Appl. Opt. **12**, 1951-1956 (1973)

3.6 H. Kohler: Zur Vermessung holographisch-interferometrischer Verformungsfelder. Optik **47**, 271-282 (1977)

3.7 N.G. Vlasov, A.E. Shtan'ko: Measurement error evaluation in the holographic interferometry of reflecting objects, in *Proc. VII All-Union School on Holography* (LINP, Leningrad 1975) pp.212-222

3.8 D. Noblis, C.M. Vest: Statistical analysis of errors in holographic interferometry. Appl. Opt. **17**, 2198-2204 (1978)

3.9 W. Osten, C. Lange: Einige ergänzende Betrachtungen zur statistischen Fehler-Analyse nach Nobis und Vest. FMC-Ser. Inst. Mech. Akad. Wiss. DDR N13, 88-102 (1984)

3.10 W. Osten: Some consideration on the statistical error analysis in holographic interferometry with application to an optimized interferometry. Opt. Acta **32**, 827-838 (1985)

3.11 W. Osten: Zur optimalen Aufnahme und Auswertung holographischer Interferogrammen unter Berücksichtigung der Möglichkeiten der digitalen Bildverarbeitung. FMC-Ser. Inst. Mech. Akad. Wiss DDR N13, 14-43 (1984)

3.12 S.T. De, A.G. Kozachok, A.V. Loginov, Yu.N. Solodkin: Holographic interferometer with a minimal deformation and displacement measurement error, in *Golograficheskiye Izmeritelnye Sistemy* (holographic measurement systems), ed. by A. Kozachok (NETI, Novosibirsk 1986) pp.30-50

3.13 V.S. Pisarev, V.P. Shchepinov, V.V. Yakovlev: Optimization of a holographic interferometer for determining displacement vector components from relative fringe orders. Opt. Spektrosk. **56**, 900-906 (1984)

3.14 V.S. Pisarev, V.V. Yakovlev, V.O. Indisov, V.P. Shchepinov: Experimental design for deformation measurement by holographic interferometry. Zhur. Tekh. Fiz. **53**, 292-300 (1983)

3.15 K. Erkler, L. Wenke, W. Schreiber: Berechnung von Hologramminterferometern bestmöglicher Konditionierung. Feingerätetechnik **29**, 510-514 (1980)

3.16 H. Kohler: Holografische Interferometrie: VI. Aspekte der Auswertungs-Optimierung. Optik **62**, 413-423 (1982)

3.17 T. Matsumoto, K. Iwata, R. Nagata: Measuring accuracy of three-dimension displacements in holographic interferometry. Appl. Opt. **12**, 961-967 (1973)

3.18 H. Kohler: General formulation of the holographic-interferometric evaluation methods. Optik **47**, 469-475 (1977)

3.19 P. Weselowski: Some practical problems of displacement strain measurement by incoherent superposition of interferograms. Mechanika Teoretyczna Istosiwana **22**, 69-76 (1984)

3.20 L. Ek, K. Biedermann: Analysis of a system for hologram interferometry with a continuously scanning reconstruction beam. Appl. Opt. **16**, 2535-2542 (1977)

3.21 O.G. Lisin: On the accuracy of measuring spatial displacements of diffusely reflecting objects by holographic interferometry. Opt. Spektrosk. **50**, 521–531 (1981)

3.22 D.K. Fadeyev, V.N. Fadeyeva: *Vychislitelnye Metody Lineinoi Algebry* (numerical methods of linear algebra) (Fizmatgiz, Moscow 1963)

3.23 V.V. Voevodin: *Vychislitelnye Osnovy Lineinoi Algebry* (numerical methods in linear algebra) (Nauka, Moscow 1977)

3.24 G. Forsythe, C.B. Moler: *Computer Solution of Linear Algebraic Systems* (Prentice-Hall, Englewood Cliffs, N.J. 1967)

3.25 J.D. Day: Errors in the computation of linear algebraic systems. Int'l J. Math. Educ. Sci. Tech. **9**, 89–95 (1978)

3.26 L. Wenke, W. Schreiber, K. Erler: Genauigkeit von Verfahren zur quantitativen Auswertung holografischer Interferogrammen. Feingerätetechnik **29**, 413–416 (1980)

3.27 A.N. Guz', I.S. Chernyshenko, V.I. Chekhov: *Tsilindricheskiye Obolochki Oslablennye Otverstiyami* (cylindrical shells weakened by holes) (Naukova Dumka, Kiev 1974)

3.28 V.O. Indisov, V.S. Pisarev, V.P. Shchepinov, V.V. Yakovlev: Using reflection hologram interferometers to study local strains. Zhur. Tekh. Fiz. **56**, 701–707 (1986)

3.29 V.O. Indisov, V.S. Pisarev, V.P. Shchepinov, V.V. Yakovlev: Comparison of holographic interferogram interpretation from absolute and relative fringe orders in strain measurement, in *Deformatsiya i Razrusheniye Materialov i Konstruktsii Atomnoi Tekhniki* (strain and fracture of materials and structures in atomic engineering) (Energoizdat, Moscow 1983) pp.45–54

3.30 R.J. Pryputniewicz, W.W. Bowley: Techniques of holographic displacement measurement: An experimental comparison. Appl. Opt. **17**, 1748–1756 (1978)

3.31 V.A. Zhilkin, S.I. Gerasimov: On a possibility of studying deformed state with an overlay interferometer. Zhur. Tekh. Fiz. **52**, 2072–2085 (1982)

3.32 V.A. Zhilkin, S.I. Gerasimov, V.N. Sandarskii: Error evaluation of displacement measurements with an overlay interferometer. Opt. Spektrosk. **62**, 1385–1389 (1987)

Chapter 4

4.1 P.M. De Larminot, R.P. Wei: A fringe-compensation technique for stress analysis by reflection holographic interferometry. Exp. Mech. **16**, 241–248 (1976)

4.2 T. Matsumoto, K. Iwata, R. Nagata: Simplified explanation of fringe formation in deformation measurements by holographic interferometry. Bull. Univ. Osaka Prefect. A22, 101–110 (1973)

4.3 S.P. Timoshenko, J.N. Goodier: *Theory of Elasticity* (McGraw-Hill, New York 1970)

4.4 C.G. Foster: Accurate measurements of Poisson's ratio in small samples. Exp. Mech. **16**, 311–315 (1976)

4.5 R. Jones, D. Bijl: A holographic interferometric study of the end effects associated with the four-point bending technique for measuring Poisson's ratio. J. Phys. E 7, 357–358 (1974)

4.6 I. Yamaguchi, H. Saito: Application of holographic interferometry to the measurement of Poisson's ratio, Jap. J. Appl. Phys. **8**, 768–771 (1969)

4.7 E.E. Rowlands, J.M. Daniel: Application of holography to anisotropic composite plates. Exp. Mech. **12**, 75–81 (1972)

4.8 I.N. Odintsev, V.P. Shchepinov, V.V. Yakovlev: Determination of elastic constants of materials by holographic interferometry. Zhur. Tekh. Fiz. **58**, 108–113 (1988)

4.9   P. Boone, R. Verbiest: Application of hologram interferometry to plate deformation and translation measurements. Opt. Acta 16, 555-567 (1979)

4.10 R.C. Sampson: Holographic interferometry applications in experimental mechanics, Exp. Mech. 10, 313-320 (1970)

4.11 A.D. Wilson, C.H. Lee, H.R. Lominac, D.H. Strope: Holographic and analytic study of semi-clamped rectangular plate supported by struts. Exp. Mech. 11, 129-234 (1971)

4.12 V.V. Yakovlev, V.P. Shchepinov, V.S. Pisarev, I.N. Odintsev: Sensitivity of holographic interferometry in strain measurement, in *Fizika i Mekhanika Deformatsii i Razrusheniya Konstruktsionnykh Materialov* (mechanics of strain and fracture in structural materials) (Atomzidat, Moscow 1978) pp.138-149

4.13 T.R. Hsu: Large deformation measurements by real-time holographic interferometry. Exp. Mech. 14, 408-411 (1974)

4.14 N.G. Vlasov, S.G. Galkin, Yu.P. Presnyakov: Improving the sensitivity of the holographic interferometry of diffusely reflecting objects, in *Metody i Ustroistva Opticheskoi Golografii* (methods and instrumentation of optical holography) (LNPI, Leningrad 1983) pp.148-162

4.15 A. Lev, I. Politch: Measuring the displacement vector by holographic interferometry. Opt. Laser Techn. 11, 45-47 (1979)

4.16 N. Abramson: Sandwich hologram interferometry: 2. Some practical calculations. Appl. Opt. 14, 981-984 (1975)

4.17 A.D. Wilson: In-plane displacement of a stressed membrane with a hole measured by holographic interferometry. Appl. Opt. 10, 908-912 (1971)

4.18 A.E. Ennos: Measurement of in-plane surface strain by hologram interferometry. J. Phys. E 1, 731-734 (1968)

4.19 P.M. Boone: Holographic determination of in-plane deformation. Opt. Technol. 2, 94-98 (May 1970)

4.20 W.J. Beranek, A.J.A. Bruinsma: A geometrical approach to holographic interferometry. Exp. Mech. 20, 289-300 (1980)

4.21 C.A. Sciammarella, J.A. Gilbert: A holographic-moiré technique to obtain separate patterns for components of displacements. Exp. Mech. 12, 1951-1956 (1974)

4.22 C.A. Sciammarella: Holographic moiré, an optical tool for the determination of displacements, strains, contours and slopes of surfaces. Opt. Eng. 21, 447-457 (1982)

4.23 T.T. Hung, C.E. Taylor: Measurement of surface displacement normal to the line of sight by holo-moiré interferometry. J. Appl. Mech. 42, 1-4 (1975)

4.24 W.J. Beranek, A.J.A. Bruinsma: Determination of displacement and strain fields using dual-beam holographic-moiré interferometry. Exp. Mech. 22, 317-323 (1982)

4.25 J.A. Gilbert, G.A. Exner: Holographic displacement analysis using image-plane techniques. Exp. Mech. 18, 382-388 (1978)

4.26 C.A. Sciammarella, S.K. Chawla: A lens holographic-moiré technique to obtain components of displacements and derivatives. Exp. Mech. 18, 373-381 (1978)

4.27 V.A. Zhilkin, L.A. Borynyak: Obtaining linear rasters and moiré effect by means of standing light waves. Izv. Vuzov, Stroit. Arkhitekt. 4, 168-170 (1975)

4.28 M.I. Marchant, S.M. Bishop: An interference technique for the measurement of in-plane displacements of opaque surfaces. Strain Anal. 9, 36-43 (1974)

4.29 V.A. Zhilkin, V.S. Zinovyev, T.V. Gorbunova: Holographic moiré study of anisotropic problems in solid mechanics. Mekh. Kompozits. Mater. 2, 341-347 (1983)

4.30 V.A. Zhilkin: Comparative analysis of the possibilities offered by holographic interferometry and holographic moiré technique in the solution of plane problems in solid mechanics. Izv. Vuzov, Stroit. i Arkhitekt. 10, 132-136 (1981)

4.31 Y. Katziz, A.A. Friesem, I. Glaser: Orthogonal in-plane and out-of-plane fringe maps in holographic interferometry. Opt. Lett. **8**, 163-165 (1983)

4.32 Y. Katziz, I. Glaser: Separation of in-plane and out-of-plane motions in holographic interometry. Appl. Opt. **21**, 678-683 (1982)

4.33 M. Yonemura: Holographic measurement of in-plane deformation using fringe visibility. Optik **63**, 167-177 (1983)

Chapter 5

5.1 P. Boone, R. Verbiest: Application of hologram interferometry to plate deformation and translation measurements. Opt. Acta **16**, 555-567 (1969)

5.2 R.C. Sampson: Holographic interferometry applications in experimental mechanics. Exp. Mech. **10**, 313-320 (1970)

5.3 V.T. Sapunov, V.G. Seleznev, V.P. Shchepinov, V.V. Yakovlev: Using holographic interferometry for studying the stressed state in solids, in *Problemy Golographii* (problems in holography) Vol.3 (MIREA, Moscow 1973) pp.29-32

5.4 A.D. Wilson, C.H. Lee, H.R. Lominac, D.H. Strope: Holographic and analytic study of semi-clamped rectangular plate supported by struts. Exp. Mech. **11**, 229-234 (1971)

5.5 I.N. Odintsev, V.P. Shchepinov, V.V. Yakovlev: Evaluation of real deformation conditions in axially symmetric plates in a holographic experiment, in *Prochnost i Dolgovechnost Materialov; Konstruktsii Atomnoi Tekhniki* (strength and durability of materials and structures in atomic engineering) (Energoizdat, Moscow 1982)

5.6 R.E. Rowlands, J.M. Daniel: Applications of holography to anisotropic composite plates. Exp. Mech. **12**, 75-82 (1972)

5.7 M.I. Marchant, M.B. Snell: Determination of the flexural stiffness of thin plates from small deflection measurements using optical holography. J. Strain Anal. **17**, 53-61 (1982)

5.8 V.A. Zhilkin: Comparative analysis of the possibilities offered by holographic interferometry and holographic moiré technique in the solution of plane problems in solid mechanics. Izv. Vuzov, Stroit. Arkhitekt. **10**, 132-136 (1981)

5.9 T.D. Dudderar, E.M. Doerries: A study of effective crack length using holographic interferometry. Exp. Mech. **16**, 300-304 (1976)

5.10 G. Frankowski: Anwendungsmöglichkeiten der holografischen Interferometrie in der experimentellen Bruchmechanik. FMC- Sec. Inst. Mech. Akad. Wiss. DDR N3. 46-57 (1982)

5.11 A.N. Grishanov, A.J. Pavlikov, V.A. Khandogin: On the noninvariance of J-integral in plastic deformation. Problemy Prochnosti **5**, 11-19 (1986)

5.12 M.C. Collins, C.E. Watterson: Surface-strain measurements on a hemispherical shell using holographic interferometry. Exp. Mech. **15**, 128-132 (1975)

5.13 T. Matsumoto, K. Imata, R. Nagata: Measurement of deformation in a cylindrical shell by holographic interferometry. Appl. Opt. **13**, 1080-1084 (1974)

5.14 C.A. Sciammarella, T.Y. Chiang: Holographic interferometry applied to the solution of shell problem. Exp. Mech. **14**, 217-224 (1974)

5.15 V.V. Yakovlev, V.P. Shchepinov, V.S. Pisarev, V.O. Indisov: Holographic interferometric study of a cylindrical shell with a rectangular cutout. Prikl. Mekh. **20**, 117-120 (1984)

5.16 V.A. Zhilkin, A.P. Ustinenko, L.A. Borynyak: Study of the deformation of thin-walled circular cylindrical shells with a panoramic interferometer. Prikl. Mekh. **12**, 79-84 (1986)

5.17 V.O. Indisov, V.P. Shchepinov, V.V. Yakovlev: Combination of holographic interferometry with the finite element method in evaluating the strength and stiffness of shells, in *Issledovaniye Prochnosti Materialov i Konstruktsii Atomnoi*

Tekhniki (studies of the strength of materials and structures in atomic engineering) (Energoizdat, Moscow 1984) pp.56-60

5.18 W. Schumann, J.P. Zürcher, D. Cuche: *Holography and Deformation Analysis*, Springer Ser. Opt. Sci., Vol.46 (Springer, Berlin, Heidelberg 1985)

5.19 L.A. Borynyak, S.I. Gerasimov, V.A. Zhilkin: Experimental techniques of interferogram recording and interpretation providing the required accuracy in strain tensor component determination. Avtometriya 1, 17-24 (1982)

5.20 S.P. Timoshenko, S. Woinowsky-Krieger: *Theory of Plates and Shells* (McGraw-Hill, New York 1959)

5.21 A.N. Guz', I.S. Chernyshenko, V.I. Chekhov: *Tsilindricheskiye Obolochki Oslablennye Otverstiyami* (cylindrical shells weakened by holes) (Nauka Dumka, Kiev 1974)

5.22 T.K. Begeev, V.I. Grishin, V.S. Pisarev: Stress study of shells with drilled holes by the finite-element method and holographic interferometry. Uch. Zapiski TSAGI, 15, 85-96 (1984)

5.23 A.K. Preiss: *Opredeleniye Napryazhenii v Obyeme Detali po Danym Izmerenii na Poverkhnosti* (stress determination in the bulk of an element from surface measurements) (Nauka, Moscow 1979)

5.24 B.M. Barishpolsky: A combined experimental and numerical method for the solution of generalized elasticity problems. Exp. Mech. 20, 345-349 (1980)

5.25 K. Chandrasekhar, K.A. Jacob: An experimental-numerical hybrid technique for three-dimensional stress analysis. Int'l J. Num. Meth. Eng. 11, 1845-1863 (1977)

5.26 A.K. Preiss, A.V. Fomin: Experimental-numerical methods in the mechanics of an elastic body. Mashinostroyeniye 2, 72-79 (1986)

5.27 G.S. Pisarenko, T.Sh. Shagdyr, V.A. Khyuvenen: Experimental-numerical methods of stress concentration determination. Problemy Prochnosti 8, 3-6 (1983)

5.28 I.M. Weathery: Integration of laser-speckle and finite element techniques of stress analysis. Exp. Mech. 25, 60-65 (1985)

5.29 G.S. Pisarenko, N.A. Stetsyuk, V.A. Khyuvenen: Application of experimental-numerical techniques to study stress concentration in a beam with holes subjected to bending. Problemy Prochnosti 10, 50-54 (1986)

5.30 O.S. Zienkiewicz: *The Finite Element Method in Engineering Science* (McGraw-Hill, New York 1971)

5.31 S.L. Croish, A.M. Starfield: *Boundary Element Methods in Solid Mechanics* (George Allen & Unwin, London 1983)

5.32 B.V. Karmugin, A.N. Beskov, D.A. Mendelson: Study of the stressed state of a high pressure electromagnetic valve. Khimicheskoye i Neftyanoe Mashinstr. 6, 7-9 (1982)

5.33 S.R. McNeill, M.A. Sutton: A development for the use of measured displacements in boundary element modelling. Eng. Anal. 2, 124-127 (1985)

5.34 I. Balas, I. Sladek, M. Drzik: Stress analysis by combination of holographic interferometry and boundary-integral method. Exp. Mech. 23, 196-202 (1983)

5.35 P.F. Papkovich: *Theory of Elasticity* (Oborongiz, Moscow 1939)

5.36 E.R. Robertson, S.D. Hovanesian, W. King: The application of holography to membrane analogy for torsion, in *Engineering Use of Coherent Optics*, ed. by E.R. Robertson (Cambridge Univ. Press, Cambridge 1976) pp.47-48

5.37 P. Boone: Visualisation of Airy functions by hologram interferometry. Optik 34, 406-420 (1972)

5.38 *Eksperimental'nye Metody Issledovaniya Deformatsii i Napryazhenii* (experimental methods of stress and strain investigation) (Naukova Dumka, Kiev 1981)

5.39 R.E. Rowlands, T. Liber, J.M. Daniel, P.G. Rose: Higher-order numerical differentiation of experimental information. Exp. Mech. 13, 105-112 (1973)

5.40 L.H. Taylor, G.B. Brandt: An error analysis of holographic strains determined by cubic splines. Exp. Mech. 12, 543-548 (1972)

5.41   G.H. Ahlberg, E.N. Nilson, G.L. Walsh: *The Theory of Splines and their Applications* (Academic, New York 1967)
5.42   R. Dändliker, B. Ineichen, F.M. Mottier: High resolution hologram interferometry by electronic phase measurement. Opt. Commun. **9**, 412-416 (1973)
5.43   R. Dändliker, B. Eliasson, B. Ineichen, F.M. Mottier: Quantitative determination of bending and torsion through holographic interferometry, in *The Engineering Uses of Holography*, ed. by E.R. Robertson, J.M. Harvey (Cambridge Univ. Press, Cambridge 1970) pp.99-117
5.44   W. Schumann, M. Dubas: *Holographic Interferometry*, Springer Ser. Opt. Sci., Vol.16 (Springer, Berlin, Heidelberg 1979)
5.45   J.C. Chamet, F. Montel: Interferometrie holographique sur objects diffusants: application de la mesure du contraste a la détermination des gradients de displacement. Rev. Phys. Appl. **12**, 603-610 (1977)
5.46   J. Ebleni, J.C. Chamet: Strain components obtained from contrast measurement of holographic fringe patterns. Appl. Opt. **16**, 2543-2545 (1977)
5.47   P.J. Pryputniewicz: Holographic strain analysis. An implementation of the fringe vector theory. Appl. Opt. **17**, 3613-3618 (1978)
5.48   K.A. Stetson: Homogeneous deformation: Determination by fringe vectors in hologram interferometry. Appl. Opt. **14**, 2256-2259 (1975)
5.49   C.A. Sciammarella, R. Narayanan: The determination of the components of the strain tensor in holographic interferometry. Exp. Mech. **24**, 257-264 (1984)
5.50   H. Saito, I. Yamaguchi, I. Nakajima: Application of holographic interferometry to mechanical experiments, in *Application of Holography*, ed. by E.S. Barkett (Plenum, New York 1971) pp.105-115
5.51   K.A. Stetson: Moiré method for determining bending moments from hologram interferometry. Opt. Laser. Technol. N2, 80-85 (1970)
5.52   L. Pirodda: Optical differentiation of geometrical patterns. Exp- Mech. **17**, 427-432 (1977)
5.53   D. Paoletti, A.D. D'Altorio: Sandwich double exposure holograms for optical differentiation of displacement patterns. Opt. Commun. **55**, 338-341 (1981)
5.54   A.G. Kozachok: *Golograficheskiye Metody Issledovaniya v Eksperimentalnoi Mekhanike* (holographic methods in experimental mechanics) (Mashinostro-yeniye, Moscow 1984)
5.55   N. Abramson: Sandwich hologram interferometry: 2. Some practical calculations. Appl. Opt. **14**, 981-984 (1975)
5.56   L. Wang, J.D. Hovanesian, Y.Y. Hung: A new fringe carrier method for the determination of displacement derivatives in hologram interferometry. Opt. Laser Eng. **5**, 109-120 (1984)
5.57   D. Paoletti, G. Schirripa-Spagnolo, A.D. D'Altorio: Sandwich hologram for displacement derivative. Opt. Commun. **56**, 325-329 (1986)
5.58   G.N. Savin: *Stress Concentration Near Holes* (GITTI, Moscow 1951)

Chapter 6

6.1   G. Wernicke, G. Frankowski: Untersuchung des Fließbeginns einsatzgehärteter und gaskarbonitrierter Stahl mit Hilfe der holografischen Interferometrie. Die Technik. **32**, 393-396 (1977)
6.2   P. Hariharan, Z.C. Hegedus: Simple multiplexing technique for double-exposure hologram interferometry. Opt. Commun. **9**, 152-155 (1973)
6.3   S.A. Novikov, V.P. Shchepinov, V.S. Aistov: Study of residual deformation development in MNLZ roller band by coherent optics, in *Dinamika i Prochnost Metallurgicheshkikh Mashin* (dynamics and strength of metallurgical machines) (VNIIMETMASH, Moscow 1984) pp.47-52

6.4   V.V. Yakovlev, V.P. Shchepinov, I.N. Odintsev: Holographic study of initial residual deformation in machine parts. Probl. Prochn. **10**, 118-120 (1979)

6.5   M.L. Khenkin, I.Kh. Lokshin: *Razmernaya Stabilnost Metallov i Splavov v Tochnom Mashinostroyenii* (size stability of metals and alloys in high-precision engineering) (Mashinostroyenie, Moscow 1974)

6.6   A.G. Rakhshtadt, E.K. Zakharov, V.G. Leshkovtsev: High-sensitivity method for determining resistance of alloys to microplastic deformation under pure bending. Zavod. Labor. **36**, 980-983 (1970)

6.7   V.A. Zhilkin, S.I. Gerasimov: Application of overlay holographic interferometry to determination of elasto-plastic and residual deformations, in *Ostatochnye Tekhnologicheskie Napryazheniya* (residual technological stresses) (Inst. Mech. Probl., ANSSSR, Moscow 1985) pp.136-141

6.8   A.I. Tselikov, B.A. Morozov, O.G. Lisin, V.S. Aistov: Application of holographic interferometry to studies of elastic and plastic deformation. Probl. Proch. **6**, 106-110 (1976)

6.9   V.P. Shchepinov, B.A. Morozov, V.S. Aistov: Separation of elastic and residual deformations by holographic interferometry, in *Optiko-Geometricheshki Metody Issledovaniya Deformatsii i Napryazhenii* (optiko-geometrical methods of studying strain and stress) (DGU, Dnepropetrovsk 1978) pp.72-74

6.10 V.P. Shchepinov, V.V. Yakovlev: Holographic study of the deformation of machine parts. Zhur. Prikl. Mekh. Tekh. Fiz. **6**, 144-147 (1979)

6.11 V.P. Shchepinov, V.S. Aistov, B.A. Morozov, A.F. Arzanov: Measurement of elastic and residual gear tooth deflection by holographic interferometry. Vestn. Mashinostr. **12**, 3-6 (1980)

6.12 V.P. Shchepinov, V.V. Yakovlev: Determination of elasto-plastic deformation components by holographic interferometry. Zhur. Tekh. Fiz. **49**, 1005-1007 (1979)

6.13 V.V. Yakovlev, V.P. Shchepinov, V.S. Pisarev: Multi-exposure holographic interferometric study of the mechanical behavior of materials, in *Fizika i Mekhanika Deformatsii i Razrusheniya* (physics and mechanics of deformation and fracture) No.6 (Atomizdat, Moscow 1979) pp.108-114

6.14 V.A. Zhilkin, V.B. Zinovyev, T.V. Gorbunova: Holographic moiré study of strained bodies. Mekh. Kompozits. Mater. **2**, 341-347 (1983)

6.15 V.V. Petrov, E.A. Larionov, L.M. Plyuta: Holographic interferometric study of the creep of parts made of composites. Mekh. Kompozits. Mater. **6**, 1117-1119 (1984)

6.16 Z.N. Tsilosani, G.L. Dalakishvili, Sh.E. Kakigashvili: Holographic interferometric study of the deformation of silicate composite material (concrete), in *Mekhanica i Tekhnologiya Kompozitsionnykh Materialov* (mechanics and technology of composite materials) (Varna 1979) pp.550-553

6.17 S.V. Aleksandrovskii, A.E. Shtan'ko: Cold fracture of cellular concrete. Beton i Zhelzobeton **9**, 42-43 (1980)

6.18 I.A. Birger: *Ostatochnye Napryazheniya* (residual stresses) (Mashgiz, Moscow 1963)

6.19 V.G. Seleznev, A.N. Arkhipov, T.V. Ibragimov: Application of holographic interferometry to residual stress determination. Zavod. Labor. **42**, 739-741 (1976)

6.20 G.A. Kurov, G.P. Polyachek, N.G. Tomson: Holographic interferometric study of macrostresses in aluminum films, in *Sbornik Nauchnykh Trudov Problem v Mikroelektronike* (collection of papers on problems in microelectrons) (MIET, Moscow 1976) pp.31-33

6.21 I.V. Kononov, V.P. Shchepinov, V.V. Yakovlev, S.Y. Byslaev, S.V. Masas: Holographic interferometric study of the effect of neutron irradiation on residual stresses in the $Si(SiO_2)$ system, in *Prochnost Dolgovechnost Materialov v Konstruktsii Atomnoi Tekhnike* (strength and durability of materials used in atomic engineering) (Energoatomizdat, Moscow 1982) pp.43-47

240

6.22 B.S. Ramprasad, T.S. Radha: A simple method for the measurement of stress in evaporated thin films by real-time holographic interferometry. Thin Solid Films **51**, 335-338 (1978)

6.23 I.A. Birger, A.I. Arkhipov: Determination of residual stresses varying along the rod length. Zavod. Labor. **43**, 345-348 (1977)

6.24 V.G. Seleznev, A.N. Arkhipov, T.V. Ibragimov: Determination of residual stresses varying along the rod length by holographic interferometry. Zavod. Labor. **43**, 1131-1134 (1977)

6.25 I. Mather: Determination of initial stress by measuring the deformation around drilled holes. Trans. ASME **56**, 249-254 (1934)

6.26 A.A. Antonov, A.I. Bobrik, V.K. Morozov, G.N. Shernihev: Residual stress determinaion by hole drilling and holographic interferometry. Mekh. Tv. Tela 2, 182-189 (1980)

6.27 L.M. Lobanov, B.S. Kasatkin, V.A. Pivtorak: Holographic interferometry to study residual welding stress. Avtomat. Svarka 3, 1-6 (1983)

6.28 A.A. Rassokha, N.N. Talalaev: Evaluation of residual stresses in welded joints between thin plates. Zavod. Labor. **48**, 74-77 (1982)

6.29 S.G. Galkin, V.K. Morozov: Experimental measurements of residual stresses by holographic interferometry, in *Ostatochnye Tekhnologicheskiye Napryazheniya* (residual technological stresses) (IPM AN SSSR, Moscow 1985) pp.116-121

6.30 G.N. Chernyshev, A.A. Antonov: Residual stress determination by hole drilling and holographic interferometry, in *Ostatochnye Tekhnologicheskiye Napryazheniya* (residual technological stresses) (IPM AN SSSR, Moscow 1985) pp.348-356

6.31 A.A. Antonov: Development of techniques and equipment for holographic testing of residual stresses in welded joints. Svar. Proizv. 12, 26-28 (1983)

6.32 L.M. Lobanov, V.A. Pivtorak, S.G. Andrushchenko: Portable holographic unit for residual stress determination, in *Ostatochnye Tekhnologicheskiye Napryazheniya* (residual technological stresses) (IPM AN SSSR, Moscow 1985) pp.215-219

6.33 C.L. Chow, C.H. Cundiff: On residual stress measurements in light truck wheels using the hole-drilling method. Exp. Mech. **25**, 54-59 (1985)

6.34 V.P. Shchepinov, B.A. Morozov, S.A. Novikov, V.S. Aistov: Contact surface studies by holographic interferometry. Zhur. Tekh. Fiz. **50**, 1926-1928 (1980)

6.35 J.T. Atkinson, M.J. Labor: Measurement of the area of real contact using holographic interferometry, *Proc. Conf. Applications of Holography and Optical Data Processing* (Pergamon, Oxford 1977)

6.36 S.P. Timoshenko, J.N. Goodier: *Theory of Elasticity* (McGraw-Hill, New York 1970)

6.37 I.S. Klimenko, V.P. Ryabukha: Image subtraction holography based on spatial filtration to reveal faults in surface microrelief. Opt. Spektrosk. **59**, 398-403 (1985)

6.38 A.V. Osintsev, Yu.I. Ostrovsky, V.P. Shchepinov, V.V. Yakovlev: Effect of contact pressure on fringe contrast in holographic interferometry. Pisma Zhur. Tekh. Fiz. **11**, 202-204 (1985)

6.39 N.I. Bezukhov: *Osnovy Teorii Uprugosti, Plastichnosti i Polzuchesti* (fundamentals of the theory of elasticity, plasticity and creep) (Vyssh. Shkola, Moscow 1968)

6.40 K.N. Petrov, Yu.P. Presnyakov: Holographic interferometry of corrosion. Opt. Spektrosk. **44**, 309-311 (1978)

6.41 R.A. Ashton, D. Slovin, H.J. Gerritsen: Interferometric holography applied to elastic stress and surface corrosion. Appl. Opt. **10**, 440-441 (1971)

6.42 V.N. Kudreev, Yu.A. Panibratsev, G.S. Safronov, A.I. Safronova, V.I. Titar: On the use of holographic interferometry to reveal microrelief defects in microelectronics. Mikroelektr. **8**, 166-171 (1979)

6.43 A.P. Dmitriev, G.V. Dreiden, A.V. Osintsev, Yu.I. Ostrovsky, V.P. Shchepinov,

M.I. Etinberg, V.V. Yakovlev: Study of cavitation erosion by correlation holographic interferometry. Preprint PTI, No.1209 (1988)

6.44 Y.I. Ostrovsky, M.M. Butusov, G.V. Ostrovskaya: *Interferometry by Holography*, Springer Ser. Opt. Sci., Vol.20 (Springer, Berlin, Heidelberg 1980)

Chapter 7

7.1 B.P. Hildebrand, K.A. Haines: Multiplewave length and multiple-source holography applied to contour generation. J. Opt. Soc. Am. **57**, 155-162 (1967)

7.2 J.R. Varner: Simplified multiple-frequency holographic contouring. Appl. Opt. **10**, 212-213 (1971)

7.3 S.T. De, A.G. Kozachok, A.V. Loginov: Two-wavelength holographic contour generation, in *Oopticheskaya Golografiya i ee Primeneniye* (optical holography and its applications) (Len. Dom. Nauch. Tekh. Prop., Leningrad 1974) pp.12-15

7.4 J.S. Zelenka, J.R. Varner: Multiple-index holographic contouring. Appl. Opt. **8**, 1431-1437 (1969)

7.5 W. Schmidt, A. Vogel, D. Preussler: Holographic contour mapping using a dye laser. Appl. Phys. **1**, 103-109 (1973)

7.6 L.O. Heflinger, R.F. Wueker: Holographic contouring via multi-frequency lasers. Appl. Phys. Lett. **15**, 28-30 (1969)

7.7 I.M. Nagibina, V.L. Kozak, T.A. Il'inskaya: Holographic interferometric study of the surface shape of macroobjects, in *Opticheskaya Golografiya i ee Primeneniya* (optical holography and its applications) (Len. Dom. Nauch. Tekh. Prop., Leningrad 1974) pp.3-5

7.8 V.M. Ginzburg, B.M. Stepanov (eds.): *Golografiya: Metody i Apparatura* (holography: Methods and instrumentation) (Sov. Radio, Moscow 1974)

7.9 P.D. Henshaw, S. Ezekiel: High resolution holographic contour generation with white light reconstruction. Opt. Commun. **12**, 39-42 (1984)

7.10 T. Tsuruta, N. Shiotake, J. Tsujiuchi, K. Matsuda: Holographic generation of contour maps of diffusely reflecting surface by using immersion method. Jpn. J. Appl. Phys. **6**, 661-662 (1967)

7.11 D.M. Bavelskii, V.S. Listovets, Yu.I. Ostrovsky: Application of holographic interferometry to power plant engineering. Energomashinostr. **8**, 21-24 (1976)

7.12 D. Groves, M.J. Labor, N. Cohen, J.T. Atkinson: A holographic technique with computer-aided analysis for the measurement of wear. J. Phys. E **13**, 741-746 (1980)

7.13 L. Marti, B. Moreno, Yu.I. Ostrovsky: Difference holographic contour generation. Pisma Zh. Tekh. Fiz. **10**, 1395-1398 (1984)

7.14 N.I. Bezukhov: *Osnovy Teorii Uprugosti, Plastichnosti i Polzuchesti* (fundamentals of the theory of elasticity, plasticity and creep) (Vysshaya Shkola, Moscow 1968)

7.15 A.M. Nizhin, I.D. Torbin, F.A. Mitina: Replicas of diffraction gratings and aspheric surfaces on a layer of polyester adhesive. Zhur. Prikl. Spektr. **11**, 327-332 (1969)

7.16 N.G. Vlasov, Yu.P. Presnyakov, E.G. Semenov: Shift interferometry to study the geometric parameters of objects. Opt. Spektr. **37**, 369-371 (1974)

7.17 J.C. Dainty (ed.): *Laser Speckle*, 2nd edn., Topics Appl. Phys., Vol.1 (Springer, Berlin, Heidelberg 1984)

7.18 M. Françon: *La Granularité Laser (Speckle) et ces Applications en Optique* (Masson, Paris 1978)

7.19 F. Gymesi, Z. Füzessy: Difference holographic interferometry. Opt. Commun. **53**, 17-22 (1985)

Chapter 8

8.1   R.L. Powell, K.A. Stetson: Interferometric analysis by wavefront reconstruction. J. Opt. Soc. Am. **55**, 1593-1598 (1965)

8.2   M.M. Butusov: Time-average holographic interferometry: recording the interference pattern. Opt. i. Spektr. **37**, 532-536 (1974)

8.3   B.A. Belgorodskii, M.M. Butusov, Yu.G. Turkevich: Vibration analysis with phase-modulated reference beam. Akust. Zhur. **17**, 445-457 (1971)

8.4   V.S. Listovets, Yu.I. Ostrovskii: Vibration analysis by holographic interferometry. Zhur. Tekh. Fiz. **44**, 1345-1373 (1974)

8.5   C.H. Ågren, K.A. Stetson: Measuring the resonances of treble-viol plate by hologram interferometry. J. Acoust. Soc. Am. **51**, 1971-1983 (1972)

8.6   R. Aprahamian, D.A. Evenson: Application of holography to dynamics: High-frequency vibration of beams. J. Appl. Mech. **37E**, 287-291 (1970)

8.7   W.F. Fagan, P. Waddell, W. McCracken: The study of vibration patterns using real-time hologram interferometry. Opt. Laser Technol. **4**, 167-172 (1972)

8.8   M.O. Fein, E.L. Green: Recording high index fringes in hologram reconstructions of vibrating plates. Appl. Opt. **7**, 1864-1865 (1968)

8.9   C.R. Hazell, A.K. Mitchell: Experimental eigen-values and mode shape for flat clamped plates. Exp. Mech. **26**, 209-216 (1986)

8.10   B.S. Hockley, J.N. Butters: Coherent photography (holography) as an aid to engineering design. J. Photogr. Sci. **18**, 16-22 (1970)

8.11   M. Lurie, M. Zambuto: A verification of holographic measurement of vibration. Appl. Opt. **7**, 2323-2325 (1968)

8.12   N.E. Molin, K.A. Stetson: Measuring combination mode vibration patterns by hologram interferometry. J. Phys. Ser. E **2**, 609-612 (1969)

8.13   A. Waddell, W. Kennedy: Pockels cell stroboscopic holography. Nov. Rev. d'Opt. **1**, 13 (1970)

8.14   A.D. Wilson, D.H. Strope: Time-average holographic interferometry of a circular plate vibrating simultaneously in two rationally related modes. J. Opt. Soc. Am. **60**, 1162-1166 (1970)

8.15   M. Zambuto, M. Lurie: Holographic measurement of general forms of motion. Appl. Opt. **9**, 2066-2072 (1970)

8.16   E. Vogt, J. Geldmacher, B. Dirr, H. Kreitlow: Hybrid vibration mode analysis of rotating turbine-blade models. Exp. Mech. **25**, 161-170 (1985)

8.17   N.E. Molin, K.A. Stetson: Fringe localization in hologram interferometry of mutually independent and dependent rotation around orthogonal, non-intersecting axes. Optik **33**, 399-422 (1971)

8.18   K.A. Stetson, R.L. Powell: Interferometric hologram evaluation and real-time vibration analysis of diffuse objects. J. Opt. Soc. Am. **55**, 1694-1695 (1965)

8.19   A.N. Zaidel, Yu.I. Ostrovsky: Holographic investigation of plasma, in Proc. VIII Int'l Conf. on Phen. in Ioniz. Gases (Vienna 1967) pp.508-516

8.20   A.N. Zaidel, P.G. Malkhasyan, G.V. Markova, Yu.I. Ostrovskii: Vibration analysis by stroboscopic holography. Zhur. Tekh. Fiz. **37**, 1824-1828 (1968)

8.21   E. Archbold, A.E. Ennos: Observation of surface vibration modes by stroboscopic hologram interferometry. Nature **217**, 842-843 (1968)

8.22   P. Shaenko, C.D. Johnson: Stroboscopic holographic interferometry. Appl. Phys. Lett. **13**, 22-24 (1968)

8.23   B. Watrasiewicz: Mechanical vibration analysis by holographic methods. Opt. Technol. **1**, 20-22 (1968)

8.24   D.S. Elinevskii, R.S. Bekbulatov, Yu.N. Shaposhnikov: Application of stroboscopic holography to vibration studies. Probl. Prochn. **5**, 95-99 (1976)

8.25   C.C. Aleksoff: Time-average holography extended. Appl. Phys. Lett. **14**, 23-25 (1969)

8.26 C.C. Aleksoff: Temporally modulated holography. Appl. Opt. **10**, 1329-1342 (1971)

8.27 B.A. Belgorodskii, M.M. Butusov, Yu.G. Turkevich: Holographic study of high frequency vibrations. Avtometriya **1**, 47-53 (1972)

8.28 F.M. Mottier: Holography of randomly moving objects. Appl. Phys. Lett. **15**, 44-45 (1979)

8.29 F.M. Mottier: Time-average holography with triangular phase modulation of reference wave. Appl. Phys. Lett. **15**, 285-287 (1969)

8.30 H.J. Caulfield: Holography of randomly moving objects. Appl. Phys. Lett. **16**, 234-235 (1970)

8.31 W.T. Catney: USA Pat. **3**, 415 (1968)

8.32 H.J. Caulfield, J.L. Harris, H.W. Hemstreet: Holography of moving objects. Proc. IEEE **55**, 1758-1762 (1967)

8.33 D. Gabor, G.W. Stroke, R. Restrick: Optical image synthesis (complex amplitude addition and subtraction) by holographic Fourier transformation. Phys. Lett. **18**, 116-118 (1965)

8.34 L.F. Collins: Difference holography. Appl. Opt. **7**, 203 (1968)

8.35 M.R. Wall: The form of holographic vibration fringes. Opt. Technol. **7**, 203-208 (1968)

8.36 K. Biedermann, N.-E. Molin: Combining hypersensitization in situ processing for time-average observation in real-time hologram interferometry. J. Phys. E **3**, 669-681 (1970)

8.37 P. Hariharan: Application of holographic subtraction to time-average hologram interferometry of vibrating objects. Appl. Opt. **12**, 143-146 (1973)

8.38 Yu.I. Ostrovsky, M.M. Butusov, G.V. Ostrovskaya: *Interferometry by Holography*, Springer Ser. Opt. Sci., Vol.20 (Springer, Berlin, Heidelberg 1980)

8.39 C.R. Hazell, C.D. Liem: Vibration analysis by interferometric fringe modulation. Nov. Rev. d'Opt. **1**, 12 (1970)

8.40 C.T. Moffat, B.M. Watrasiewicz: An extention of amplitude range for time-average holography. Nov. Rev. d'Opt. **1**, 12 (1970)

8.41 M.M. Butusov, V.Ya. Demchenko, Yu.G. Turkevich: Holographic stroboscope based on passively modulated ruby laser. Prib. Tekh. Eksp. **2**, 203-204 (1971)

8.42 F.M. Mottier: Holography of randomly moving objects. Appl. Phys. Lett. **15**, 44-45 (1979)

8.43 J.P. Waters: Object motion compensation by speckle reference beam holography. Appl. Opt. **11**, 630-636 (1972)

8.44 D.S. Elenevskii, A.E. Petrenko, Yu.N. Shaposhnikov, V.A. Sharonov: Vibration study of rotating objects by synchronizing laser operation, in *Primenenie Lazerov v Nauke i Tekhnike* (laser applications in science and technology) (MIASS 1987) p.10

8.45 S. Kawase, T. Honda, J. Tsujiuchi: Measurement of elastic deformation of rotating objects by using holographic interferometry. Opt. Commun. **16**, 96-98 (1976)

8.46 Y. Chen: Holographic interferometry applied to rotating disks. Trans. ASME **42**, 499-512 (1975)

8.47 J.P. Sikora, F.T. Mendenhall: Holographic vibration study of a rotating propellor blade. Exp. Mech. **14**, 230-232 (1974)

8.48 H. Bjelkhagen, N. Abramson: Pulsed sandwich holography: Pt.I. Appl. Opt. **16**, 1727-1732 (1977)

8.49 J.C. McBain, I.E. Horner, W.A. Stange: Vibration analysis of a spinning disk using image-derotated holographic interferometry. Exp. Mex. **19**, 17-22 (1979)

8.50 J.C. McBain, W.A. Stange, K.G. Harding: Real-time response of a rotating disk using image-derotated holographic interferometry. Exp. Mech. **21**, 34-40 (1981)

8.51 K.A. Stetson: The use of an image derotator in hologram interferometry and speckle photography of rotating objects. Exp. Mech. **18**, 67-73 (1978)

8.52   T. Tsuruta, Y. Itoh: Holographic interferometry for rotating subjects. Appl. Phys. Lett. 17, 85–87 (1970)
8.53   N.V. Morozov, J.P. Alum, Yu.I. Ostrovsky: Opposed-beam holographic interferometry of rotating objects. Zhur. Tekh. Fiz. 51, 355–361 (1981)

Subject Index